Frederick S. Gay

About the Author

TIMOTHY GAY played football at Caltech and earned his Ph.D. in atomic physics from the University of Chicago. He has been a professor of physics at the University of Nebraska-Lincoln since 1993. He is a fellow of the American Physical Society and heads a research group that studies electron and neutrino physics. His video segments on football and physics have been profiled in the *Wall Street Journal, ESPN The Magazine,* and *People* magazine, and on *ABC World News Tonight,* NPR, and elsewhere.

The Physics
OF FOOTBALL

DISCOVER THE SCIENCE OF BONE-CRUNCHING HITS, SOARING FIELD GOALS, AND AWE-INSPIRING PASSES

TIMOTHY GAY, Ph.D.

With a Foreword by
BILL BELICHICK

itbooks

AN IMPRINT OF HARPERCOLLINS PUBLISHERS

TO WILLY, WHO RECRUITED ME

Photographs © by Pittsburgh Steelers Image Library (Immaculate Reception); Bettmann/CORBIS (Dick Butkus); Manny Rubio/NFL Photos (Anthony Muñoz); Walter Iooss Jr./Sports Illustrated (Jerry Rice); Jonathan Daniel/Getty Images (Barry Sanders); 2004 Rich Clarkson and Associates (Invesco Field); Associated Press/AP WideWorld (Tom Dempsey); Rob Tringali/SportsChrome (Ray Guy); E. Braverman/Getty Images (Foot kicking ball); Bettmann/CORBIS (Terry Bradshaw); Associated Press/AP WideWorld (Larry Csonka); John Zimmerman/Sports Illustrated (Chuck Bednarik/Frank Gifford); and Richard Hamilton Smith/CORBIS (Fans doing the Wave).

Photograph of Giants Stadium (page 114) is courtesy of the Giants.

A hardcover edition of this book was published in 2004 by Rodale Inc. under the title *Football Physics*.

HarperCollins books may be purchased for educational, business, or sales promotional use. For information, please e-mail the Special Markets Department at SPsales@harpercollins.com.

First Harper paperback published 2005.

Designed and diagrams by Christopher Rhoads
Charts by Sandy Freeman
Illustrations on pages 41 and 42 by Jason Schneider

Library of Congress Cataloging-in-Publication Data

Gay, Timothy J. (Timothy James).
    [Football physics]
    The physics of football : discover the science of bone-crunching hits, soaring field goals, and awe-inspiring passes / Timothy Gay ; foreword by Bill Belichick.
        p.   cm.
    Originally published: Football physics. Emmaus, Pa. : Rodale, c2004.
    ISBN-10: 0-06-082634-7
    ISBN-13: 978-0-06-082634-5
    1. Football.   2. Physics.   I. Title.
    GV959.G39   2005
    796.332'03—dc22   2005046284

    HB 12.12.2017

CONTENTS

CONTENTS

As the father of physics, Isaac Newton changed our understanding
of the world. His laboratory at Cambridge was a model of scholarly isolation, and his dedication to his work was complete. But
had Newton worked at a 21st century American university that
fielded a football team, he would have been able to get out in the
sun and observe firsthand how the laws he discovered affected
play. That's what Tim Gay does in *Football Physics*, in engaging
fashion.

The action that happens on a football field involves mass, velocity, acceleration, torque, and many other concepts covered in
this book. While some observers see only carnage and chaos, brilliant athletic performances and bone-jarring collisions, the sci-
ence-minded see the field as a working laboratory. This kind of fan
will notice that gifted athletes who play the game at the highest
level can seemingly push physics to the limit. I have had the pleasure of coaching such players in the NFL. Lawrence Taylor, whom
Tim discusses at length, personified speed and power. Phil Simms
and Tom Brady could throw consistently with the accuracy that
Tim describes in his chapter on the West Coast offense, and Dave
Jennings and Sean Landeta could punt as well as anyone in windy
Giants Stadium.

But do real players and real coaches actually use this physics
stuff during a game? To answer that question, let's consider Troy
Brown, an outstanding punt returner with the Patriots. Troy has
to stand back there and wait for the ball, knowing that the opponent's punt team is bearing down on him. If he catches the punt
cleanly, he may have a chance to break a big return. If he juggles

the football—or the "prolate spheroid," as Tim points out it's technically called—he may get broken up by the punt coverage.

The difference between these two results is the returner's ability to read the ball in flight. First he has to consider the wind conditions (returners are key players on windy days). Then he has to accurately gauge the flight of the ball: If a right-footed punter is kicking, the ball will most likely have a counterclockwise rotation as it approaches. If the punt turns over and descends with the nose down, the ball will break to the left and the returner will have to move quickly in that direction in order to be in position to catch the ball. If the ball is coming down with the nose up, it will break right. A wobbling or end-over-end punt will be short. Punt returners, the good ones like Troy, have learned and internalized these subtle physics lessons through countless hours of practice and game experience. After reading this book, you'll understand what they know by instinct.

A coach uses physics too. When we prepare for the annual NFL draft, we look at each player's speed and his quickness. This book provides the best discussion of the difference between the two that I've seen.

Whether your primary interest is popular science or pro football, you will enjoy *Football Physics*. You'll learn something about the way the world works, and you will come away with an increased appreciation for some fine points of a great game.

Bill Belichick
Foxboro, Massachusetts
April 2004

PREFACE

The largest physics class ever held was the one conducted by *Apollo 15* commander David R. Scott on August 2, 1971, while standing on the surface of the moon. Playing to a camera mounted on the lunar rover, astronaut Scott demonstrated, for everyone on earth to see (everyone with a television, anyway), that a hammer and a feather fall at exactly the same speed in the absence of air, thus proving one of Galileo's theories about objects in motion. That was a great lesson, to be sure, but then Scott's class met only once—my physics class at the University of Nebraska–Lincoln meets much more regularly.

This book was inspired by a series of lessons that I have presented, like Mr. Scott, outside of the classroom and to quite a large audience: the 78,000 fans who come to see our football team, the Nebraska Cornhuskers, play on their home turf on Saturday afternoons each fall. Back in 1999, Jeff Schmahl, the director of Huskervision, was sitting around brainstorming with a student assistant of his, Kay Dowd. Huskervision is the audiovisual part of the athletic department at the University of Nebraska; it is responsible for the programming that appears on the giant video screens at either end of Tom Osborne Field in Memorial Stadium. Jeff and Kay were trying to figure out something that might have more of an "academic flavor," as Jeff later put it to me, than the string of instant replays and player features they typically showed. Sports and science seemed like a good combination, and physics was the branch of science that seemed to pertain best to the game of football.

Jeff sent Kay over to the physics department, where she quickly ferreted out the people who are really in charge of any academic unit: the departmental secretaries. Finding them all assembled in our front office, she asked if they knew of any physics professors who loved football and were shameless self-promoters. Of course, all four of them responded in unison, "That would be Tim Gay."

Kay didn't have to sell the idea very hard to me. I've always loved football and talking with people about physics. Some of my earliest memories of playing sports with my dad were the touch-football games with the neighborhood kids in a field at the end of our street on Thanksgiving morning. Unfortunately, the public high school I went to in Ohio was too small to have a football team, and at Phillips Academy in Andover, Massachusetts, where I spent my last two years of high school, I wasn't good enough to make the varsity football team. Instead, I became the team's manager, and that gave me my first exposure to "real" football.

Andover had a very strong athletics program. The team's coach was a great guy named Steve Sorota who had played football with Vince Lombardi at Fordham University. My teammates at Andover during my senior year included the likes of Bill Belichick, the head coach of the three-time world champion New England Patriots; Ernie Adams, an assistant coach with New England who is now referred to as the Patriots' strategic "secret weapon"; and Milt Holt, who ended up as Harvard's starting quarterback. As for Coach Belichick, he was something of a football guru even as a postgraduate student at Andover. On a couple of occasions when we were in a tough spot, I remember Coach Sorota sidling over to Bill, having a brief conversation, nodding, and going back to his place on the sidelines. We won the New England Prep championship that year.

As manager, I got to see what goes on behind the scenes for a football team to be successful. As a science geek, I was particularly interested in the technical aspects of the equipment the players used: the engineering that went into their shoulder pads and helmets, the cleat design of their athletic shoes, the effect that the ball's shape had on the way the game was played—those kinds of things. I began to realize how important science and technology were to the modern game.

When I arrived at the California Institute of Technology campus the next fall, it was obvious that the school didn't offer athletic scholarships. Most of my classmates had pocket protectors, thick glasses, and a serious lack of interest in sports. But there *was* a football team. In fact, I later found out that during and immediately after World War II there had been a large number of relatively normal men at the Institute, and that the football team had actually been scouted on occasion by the pros. This was back when Caltech home games were played in the Rose Bowl. By the time I got there we were playing out behind the old gym on California Boulevard against such football powerhouses as LaVerne College, Azusa Pacific College, the University of California–Riverside (freshmen), and the La Cañada Ducks, a semi-semi-pro club team made up of retired public servants.

The good news was that this meant I was actually good enough to play on the Caltech team. I was recruited by another freshman, Willy Moss, who was one of the team's wide receivers. Willy was a big bruiser standing about 5 feet, 7 inches tall and weighing in at an intimidating 114 pounds. He introduced me to Coach Gutman, and the rest, as they say, is history. A typical season's record for Caltech during the years that I played there was 1–7. We were so bad that in 1974 the *Wall Street Journal* profiled us in their humor column on the front page. I've suppressed my recollection of most of that article, but I do remember a line about our IQs being higher than our weights, and another one about our quarterback's pass "coming down the field like a sack of wet laundry." Whenever we won, the student body would celebrate by setting up a big bonfire at the corner of Lake Avenue and Colorado Boulevard. This tradition stopped abruptly when the FBI paid a stern visit to the dean of students one year; Colorado Boulevard, the old Route 66, was a federal highway and the fires, though infrequent, were alarming motorists.

I share these embarrassing anecdotes to point out that my history with football, while not illustrious, goes back a ways. I have played the game. And I would say in all seriousness that playing football on that woeful Caltech team was the single best experience I had in college, aside from meeting my future wife.

Fast forward to 1999: When Kay Dowd came to talk to me about developing the Football Physics series at Nebraska, I had three good reasons for signing on to the project. First and foremost, I love physics and jump at any chance I can get to talk about it. Second, I love football. The third reason is less obvious but just as important. In the increasingly technological society in which we live, we need all the science education we can get. The problem is, a lot of people have science phobia. One good way to teach science is to relate it to something people actually care about. And what a lot of people, especially Nebraskans, care about is football.

At first I was a bit nervous about how Cornhusker fans would react to physics lessons being foisted upon them during a game. I didn't need to worry. Nebraska fans are the best fans in the country (or maybe just the most polite). Their attitude is a lot like those of good coaches: *If it has anything to do with football, I'm interested!* The players that we feature in the lessons are interested too. They're willing to learn anything if they think it will help them win. One of the Nebraska players most interested in helping me out with these segments was 2001 Heisman Trophy winner Eric Crouch.

Then in the fall of 2001 I was contacted by Brad Minerd at NFL Films to see if I was interested in doing the same kind of segment for them, but in a somewhat longer format and using NFL players as guinea pigs. For a naturally long-winded professor, the idea of being able to spend 5 whole minutes on a topic instead of the usual 1 minute was very attractive, and I agreed to make 21 pieces with them on the physics behind football. The segments we came up with were shown on the NFL-produced television show *Blast!*, which spreads the gospel of American football in hundreds of countries around the globe. NFL Films has made great contributions to the sport, and it has been a real honor and pleasure to work with those guys.

I have gotten a lot of feedback on both the Huskervision and the NFL Films segments. Most of it is constructive critique, with the occasional flaming e-mail about me sullying the game's purity with science. My favorite comment so far, though, came from a lady in Nebraska who asked for a segment on the physics of twirling a baton. I thanked her for the idea and said I would discuss it with the people at Huskervision. Just as she was about to hang up, she said, "You know, it's great for the young kids that go to these games to see your pieces. A lot of them probably hate science, and you're making it interesting for them. It's like you're bringing physics out of the den of nerds and into the light of day."

This resonated with me. My office and labs are located in the sub-basement of the physics building, two floors below ground level, and I have often suspected that our university administrators want to keep us atomic physicists as far away as possible from the general public.

Well, it's true. I really do work in a den of nerds. But even physicists can stick their noses aboveground occasionally.

The Physics

of FOOTBALL

"STOP THINKING, START LOSING."

—JOE MONTANA

THE STEELERS GET LUCKY

Oakland quarterback Darryl Lamonica was wrung out with the flu and exhausted from evading the likes of Joe Greene and Dwight White all day. He was clearly on his last legs when Coach Madden pulled him and sent in Kenny "The Snake" Stabler. The Raiders were down by three, and the Snake would just have to figure out a way to score on a stingy Steelers defense. Failure was not an option—this was the first AFC divisional play-off game for both teams, and the loser's season was over. Stabler, Oakland's third-string quarterback, took the snap and immediately fumbled the ball. Pittsburgh, smelling its first postseason victory in more than 40 years, capitalized on the turnover by scoring on a Roy Gerela field goal. The crazed crowd at Three Rivers Stadium roared and thrashed the air with their Terrible Towels.

Following the kickoff, though, Stabler began to move the ball. Starting from his own 20-yard line, he quickly and coolly took his team down to the Pittsburgh 30. Then, with a first down and 10 yards to go, he dropped back to pass and realized that the Steelers' safeties were blitzing—there was running room along the left sideline. The Snake bootlegged left and took the ball in for the Raiders' first score of the day. A George Blanda extra point made it 7–6 Oakland, with 1:13 left in the game.

Steelers quarterback Terry Bradshaw had been having, if it were possible, a worse day than Lamonica. The Raiders safeties had been wreaking havoc on his receivers; he had completed just 10 of 24 pass attempts for 115 yards. Now he had to move the ball from his own 20 to a point where Gerela had some chance of kicking the winning field goal. Things, though desperate, started off well. Bradshaw had an arm like a shotgun, especially at close range, and he completed two passes in a row. At his own 40-yard line, with the game clock winding down, Bradshaw threw three straight incompletions into the same kind of coverage he had been facing all day. Miraculously, he wasn't intercepted.

It was now fourth down and 10 with 26 seconds left to play. The Steelers' new head coach, Chuck Noll, who had been sending in the plays from the sidelines, called for a short pass up the middle to rookie receiver Barry Pearson. Bradshaw took the snap and set up in the pocket to pass, but was immediately flushed out to the right by defensive tackle Art Thoms. About to be sacked, Bradshaw spotted a flash of black and gold downfield straight ahead, almost obliterated by the sea of black-and-silver defenders. He shot a tight spiral directly at the target and was immediately buried by the Raiders' defensive line.

Bradshaw's desperate strike was rocketing toward his running back, John "Frenchy" Fuqua. Fuqua had just run a buttonhook pattern and was momentarily open at the Raiders' 35. Trying to avoid his coverage, he spotted the ball just after Bradshaw released it. So did Oakland's free safety, Jack Tatum, a savage tackler known affectionately throughout the league as the Assassin. Tatum dropped backed a few steps to give himself running room to build up a bit more momentum when he stuck his helmet in the Frenchman's back, and to time the hit so that it would coincide with the arrival of the ball. It did. Tatum plowed into Fuqua with a sledgehammer tackle that sent both Frenchy and the ball flying. Tatum had achieved his goal of preventing a reception by the running back. He stood over Fuqua, grinning.

But suddenly the Raiders realized they had a problem. The ball, whether it had ricocheted off Fuqua or Tatum, was flying backward to the Raiders' 42, where Franco Harris, another rookie and the Steelers' other running back, happened to be. Getting his hands under the ball just inches above the Astroturf, Harris scooped it up and, accelerating past a group of befuddled Raiders, took it in for the winning touchdown.

Not only were the Raiders confused, so were the referees. In 1972, NFL rules stated that once a thrown ball came in contact with another offensive player, it could not subsequently be caught by a third. Thus, if Fuqua had touched the ball, the play would be ruled as an incomplete pass. Tatum's grin changed to a look of disbelief as Harris blew past him. He spun around to confront his opponent on the ground.

"Tell them you touched it, Frenchy! Tell them you touched it!"

The Frenchman never did. The referees ruled in favor of Pittsburgh. And to many people, that play—which almost immediately became known as the Immaculate Reception—marked the beginning of the Steelers dynasty of the 1970s. To this day, Frenchy Fuqua isn't talking about whether he touched the ball or not. But what almost everyone agrees on is this: It was one of the greatest plays in the history of American football.

Raiders' safety Jack "The Assassin" Tatum levels the Steelers' Frenchy Fuqua to put into motion what many consider to be the most memorable play in pro football history: the Immaculate Reception. Note the ball being deflected in the general direction of number 32 (in the background), rookie running back Franco Harris, who would scoop up the ball just inches from the turf and race 42 yards for the game-winning touchdown.

THE PHYSICS OF FOOTBALL

The haphazard beauty of plays like the Immaculate Reception is what makes football, in my opinion, the most exciting, interesting, and complex sport ever invented. And all of it—the extreme pressure and huge stakes, Bradshaw's desperate scramble to avoid being sacked, a determined defensive pursuit, Bradshaw's bullet pass to Fuqua, the freight-train collision between Tatum and Fuqua, the ball's bizarre trajectory after caroming off Tatum (or was it Fuqua?!), and, dramatically, the elegant sweeping up of the ball and run by Harris to the end zone and victory—all of it, as gut-wrenching (if you were a Raiders fan) and amazing (if you were rooting for the Steelers) as it seemed at the time and still seems when the Reception is inevitably invoked before every Super Bowl, was completely directed by the laws of physics.

For example, the fact that both teams struggled on offense throughout the game was due largely to the condition of the artificial turf, part of which had frozen solid in the cold Pittsburgh air the night before, making footing difficult for receivers and running backs alike—illustrating the impact of frictional forces, or rather the lack thereof. The war being fought on the line of scrimmage, and in particular the defensive line play, illustrated the huge energy dissipation from thousands of pounds of mass colliding on each snap and was another factor that kept the score low and the game close to the very end. Bradshaw's pass on the play was swift and accurate, hitting its target—as did Tatum!—and illustrating the effects of angular momentum and impulse. Tatum's hit on Fuqua demonstrated the concept of elastic collisions and conservation of momentum. And Harris's lucky reception and his open-field run to the goal line was a textbook example of the physics concepts of acceleration, velocity, and two-dimensional motion.

And yet, unless you're an aspiring physicist, you might well ask yourself: But what exactly *is* physics, and, more important, why should I care?

Physics is the science that describes nature at its most fundamental level. It deals with forces, matter, and energy and the interactions between them. Chemistry, biology, and geology all boil down to the underlying physics of atoms and molecules and the forces that govern them. As for how physics can illuminate the game of football, a personal anecdote might help explain.

Back when I was a graduate student at the University of Chicago, I had a lot of friends who played soccer every weekend out on the Midway. Back then, soccer seemed stupid to me—guys running around aimlessly, flailing away at the ball without using their hands, rarely scoring. How boring! Growing up in rural Ohio in the 1960s, my buddies and I referred to soccer as Commie football; there seemed to be something vaguely anti-American about it. Not knowing much about the game, I derided it as being inferior to *real* football.

Over the course of several years, though, my friends instructed me on soccer's finer points: the art of dribbling; the strategic importance of passing the ball; how to make the ball curve where you want it to go. Now don't get me wrong. I *still* think soccer is inherently inferior to football. But I can enjoy watching professional soccer matches now. My point is simply this: The more you understand about football—or any other sport, or art, or music, for that matter—the more you'll get out of watching or playing it. And to really appreciate the game of football at the deepest level, you've got to have an understanding of the basic physics principles that underlie the game. I'm not saying you can't enjoy or play the game if you don't know any physics—heaven knows if Artie Donovan knew the difference between a kilogram and a keister. What I'm saying is that by learning some simple physics, you'll enjoy the game even more than you do already.

Good players and coaches know how important basic football techniques—aka the fundamentals—are to individual performance and to the ultimate success of the team. As the game of football evolves and becomes ever more complex, a basic understanding of the physics behind it becomes even more important. A lineman playing in the 1882 Yale vs. Harvard game could rely on pure athleticism to gain the upper hand against his opponent on the other side of the line. These days, a quarterback running a pro set West Coast offense had better have athleticism coming out of his ears, but he'd also gain an edge by having a good working knowledge of kinematics.

So I'm aiming this book at fans, players, and coaches who love football so much that they're even willing to tackle some physics to increase their enjoyment of the game. It is my sincere hope that after a while they'll start to find the physics as much fun as I do.

SOME BASICS

Galileo, whose passion for understanding the natural world helped bring science to the masses, believed that "the book of nature is written in the language of mathematics." Unfortunately,

I've seen a lot of young people come out of high school with mixed feelings about science and a real fear of math. My job as a professor is to make physics interesting, and in this book I try to keep things simple enough so you can follow along as I share some observations about the remarkable, unseen laws that govern football.

As you flip through the pages of *The Physics of Football* you may notice, along with the neat photographs, some graphs and figures. Don't be alarmed! Graphs and figures are just pictures that can help explain a concept; after all, a picture is worth a thousand words, right? You also might have spotted some scary-looking math equations lurking here and there in the text. I have done my best to keep the math in this book to a minimum. But why, you might ask, use it at all? Imagine that you want to learn to drive a car and that I am your driving instructor. My job would be teaching you how to start the car, pull out into traffic, shift gears, avoid hitting other cars—everything that driving entails. What if I never spoke to you but instead communicated by motions and grunts (in much the same way my two teenage boys communicate with me)—you spoke English, I spoke English, but we didn't talk. That wouldn't make any sense. I could communicate much more effectively by using a few carefully chosen words. The same holds true if I'm trying to teach you some simple physics ideas. I'll communicate most of the concepts in this book through English and picture graphs, but occasionally it'll be a lot simpler and more effective to use basic math. At those times, we'll just grit our teeth and do it—no pain, no gain, as athletes say.

The equations we'll use are all in the form $A = B \times C$, where A, B, and C are algebraic symbols representing some physical quantity, such as the mass of William "The Refrigerator" Perry or the distance on the field, in feet, from Ty Law to Torry Holt. One common equation we'll be using, for example, is Newton's Second Law, $F = ma$, which establishes that force equals mass times acceleration.

Occasionally we'll use Greek letters instead of the usual Latin ones as symbols. This is not to make the equations more obscure. The reason is that once we run out of good Latin symbols we have to start using other alphabets. Sometimes you'll see a superscript number hanging over an equation, up in the air. For example, we might write $a = \omega^2$. This looks pretty intimidating, until you realize that the Greek letter omega is just a symbol for some physical quantity and the superscript simply means "squared": $\omega^2 = \omega \times \omega$.

MIND YOUR UNITS

When we use equations to relate physical quantities like length or weight, we have to make sure that the same quantity is represented on both sides of the equation. In other words, we can't equate apples to oranges. We'll see how this works in more detail in chapter 1, but I raise the issue now because we have to talk about units of physical quantities.

In physics, we use systems of units to describe all possible physical quantities. The two most popular systems of units are the metric system and the English system. The metric system uses the meter as the unit of length, the kilogram as the unit of mass, and the second as the unit of time. For this reason it is often called the MKS system for short. The English system is based on the foot for length, the pound for force, and the second for time—the FPS system for short. As you might have guessed, it was the English who came up with the English system, and it is the one used in American football. It was the French who came up with the MKS system, and it, of course, is used in soccer.

Our challenge here is that scientists prefer the MKS system too, for three reasons. First, physics is an international discipline, and more countries around the world use the MKS system than the English system. (Another name for the MKS system is the *Système Internationale d'unités*.) Second, humans use a base-10 number system; that is, one in which there are 10 symbols for numbers: 0, 1, 2, 3, 4, 5, 6, 7, 8, and 9. This evolved naturally from the fact that we have 10 fingers. The MKS system also uses a base-10 system: 10 millimeters equals a centimeter, 10 centimeters equals a decimeter, 10 decimeters equals a meter, etc. Thus, MKS units are naturally counted out on our fingers. (During my football career I noticed that my fellow linemen counted on their fingers as a matter of course.) Finally, units in the MKS system are highly interconnected. For example, the MKS unit of volume, the liter, is 1,000 cubic centimeters. Nothing is connected simply like this in the English system—a gallon is not directly related to a foot.

The English system does have a few things going for it. For one, the 12-inch foot divides naturally by 2, 3, 4, and 6, as opposed to just 2 and 5 for the meter. This just means that there are more ways of naturally dividing up the foot than there are the centimeter. But the most important advantage to using the English system for this book is that it is traditionally what has been used in football. We talk about yard lines, not meter lines. We're in awe of a defensive lineman who weighs 380 pounds, not 173 kilograms. Since this is a book about football, we'll use English units.

NEWTON IS THE MAN

Walter Camp (1859–1925) was more responsible than anyone else for the development of the modern game of American football. In the same way, Sir Isaac Newton essentially invented physics; or perhaps a better way to put it is that God invented physics and Newton just discovered it. (That God invented football goes without saying.)

The kind of physics that we will use in this book is known as classical physics, or Newtonian physics. Before Newton, who lived from 1642 to 1727, physics as a coherent body of knowledge didn't really exist. There were no fundamental "laws" that explained the results of the few experimental investigations that had been carried out up to that point.

Newton's chief predecessors were Aristotle, Tycho Brahe, Johannes Kepler, and Galileo. Aristotle laid the foundation of all modern science by insisting that nature operated according to underlying principles that could be understood in a systematic way.

He assumed that these principles could be figured out by brain-power alone, without actually doing any messy experiments. Big mistake. It was Galileo and Brahe who made the first careful, orderly investigations of the mechanical universe—they did the first real physics experiments. Brahe made highly accurate measurements of the motion of the planets. Kepler took these observations and showed that they yielded regular, predictable planetary motion. Galileo was concerned with the way in which earthly objects moved under the influence of forces that acted upon them, and how their position, speed, and change of speed with time were related. None of these guys, however, ultimately was able to figure out the laws underlying his observations.

This took a genius of Newton's caliber. He discovered the mathematical laws that explained all of Galileo and Brahe's observations. This was particularly impressive given that Brahe's data came from systems that were literally on an astronomical scale, while Galileo's studies had been done mostly on the top of a table. Perhaps even more impressive was the fact that Newton had to invent the mathematics (calculus) to do all this. Oh, and in his spare time he invented the modern bridge and the modern telescope.

Newtonian physics was believed to be God's Truth until the latter part of the nineteenth century when, to the consternation of physicists of the day, new experimental facts began to rear their ugly heads. They showed that Newton's ideas, while not completely wrong, did not accurately describe nature on a very small scale or at very high speeds. To do that took "modern" physics—the new theories of quantum mechanics and relativity. Relativity theory becomes important at very high speeds; say, 10,000 miles per second. Quantum mechanics comes into play when we want to talk about the movement of submicroscopic particles such as atoms and molecules. Rest assured that for things the size of footballs and football players, Newtonian mechanics is completely adequate.

PHYSICS VS. ABILITY

Finally, before we kick off, I want to address the limits of what we can learn about football by studying physics. I get asked a lot of questions that go something like this: "Which team does physics predict will win on Sunday, the Bills or the Chargers?" or "What is the physics of why Emmitt Smith is a better running back than so-and-so?" or "What does physics tell us about why safety Mike Minter never gets injured?" These questions are tough to answer because physics doesn't tell us much that is useful in these situations.

It is certainly true that neither Emmitt Smith nor Michael Vick does anything out there on the field that is forbidden by the laws of physics. Not one player in the history of the NFL (or even the XFL!) has ever failed to conserve energy or momentum, kicked a ball 300 yards, or hit another player harder than that player hit him. Physical law places absolute limits on what players can and can't do. We *can* use physics to understand why the tried-and-true, basic advice that coaches give to their players about technique works so well in football. We can use physics to reveal just how incredibly talented NFL athletes have to be to do what they do, and in such spectacular fashion. But when we get into the detailed differences between the running ability of two players, for example, or try to analyze why a bad team occasionally whips a good one, it becomes increasingly difficult to make definitive statements.

Part of the problem is that human beings are extremely complicated biomechanical machines. The attempt to make a detailed analysis of how humans move, especially with regard to sports activities, is the province of *kinesiology*. This word has the same root as *kinematics*, which is the branch of physics that studies motion. One of the main goals of kinesiology is to develop guidelines for what is and isn't good technique in a given sports activity. The problem is, a physicist can tell you how much force Jack Tatum exerted on Frenchy Fuqua during his historic hit back in 1972, and a kinesiologist can tell you which muscle groups Franco Harris used when he reached down to make his game-winning grab, but even in retrospect neither could tell Terry Bradshaw exactly how to evade all those charging Raiders pass rushers just (barely) long enough to get off his desperate but immortal toss. With due respect to Bradshaw's offensive line, that particular question escapes scientific characterization, as do intangibles such as instinct, heart, and an enduring will to win—all of which make the game of football as great as it is.

So what *can* physics tell us? Physics can tell a coach that tall placekickers, on average, will put the ball in the end zone more reliably than short placekickers. This is because the longer legs of the taller players have a larger moment of inertia and will thus give the kicked ball a larger launch speed (as you'll see in chapter 5). Physics won't explain why one 6-foot-2 placekicker with good technique is better than another 6-foot-2 kicker with seemingly equivalent technique—now we're talking talent and physiology—but it can tell the coach about general tendencies that he should look for in kickers.

Here's another example to consider. When a pro coach is looking for wide receivers to draft, should he always pick the guy with the best time in the 40-yard dash? Well, it depends. Kinematics tells us that 40-time is determined primarily by a player's ability to accelerate off the line and not by his top-end speed. Thus, if your playbook calls for a lot of short passing routes, 40-time is crucial. On the other hand, to go deep you'll want a guy with a high top-end speed like Jerry Rice, who can blow past a free safety. Such a receiver may or may not have an awesome 40-speed.

Oh, and there's one more thing physics can tell us: the referees made the right call in the Immaculate Reception. But that will have to wait until the end of the first chapter. Enough with the preliminaries. Let's see if we can discover something new about football. Ready . . . break!

"SOME PEOPLE TRY TO FIND THINGS IN THIS GAME THAT DON'T EXIST, BUT FOOTBALL IS ONLY TWO THINGS— BLOCKING AND TACKLING."

—VINCE LOMBARDI

CHAPTER 1

BLOCKING AND TACKLING

In football, there is a name for teams that fail to execute the fundamentals: losers. And there is nothing more fundamental in football than blocking and tackling. Incredible catches and jaw-dropping runs are fun to watch and can make the difference in a game or two over the course of a season, but they ultimately mean little if the team is failing at the basics.

A superb athlete who has been well coached and has the aggressive desire to make an impact on the game will consistently make solid tackles and blocks—the kind that make his team's plays work and force those of his opponents to fail. A classic example of this kind of tackler is Chicago Bears middle linebacker Dick Butkus. Butkus, a Hall of Famer who played in eight straight pro bowls, had the work ethic and technical skills of a consummate professional along with the heart of a warrior. Hanging back from the line of scrimmage to get a feel for where a play was going, he would close with ferocious speed on a runner as soon as he saw him hitting a gap in the line. Looming in the runner's path with his head up and shoulders squared, he would drive through the ball carrier with an incredibly beautiful, fluid motion that often resulted in the ball coming loose and the runner lying flat on his back and driven into the turf.

The organized mayhem that occurs on a football field is governed by the fundamental laws of classical physics: Sir Isaac Newton's three laws of motion. By breaking down blocks and tackles using these laws, we can appreciate how they *should* be

The Bears' ferocious Dick Butkus, number 51, wraps up the entire Green Bay front line—and makes the tackle.

made while discovering some truly astounding aspects of such collisions. An example of this is the force that two players exert on each other during a big hit.

FIRST, NEWTON'S FIRST LAW

We begin to gain insight into the mechanics of football collisions by considering Newton's First Law. This is the one with which you are perhaps most familiar. It says simply that mass wants to continue doing what it's already doing, whether it's at rest or in motion.

While this seems pretty straightforward, the First Law is actually counterintuitive. When you think about it, our everyday experience suggests that objects *want* to come to a stop. If you are driving along on a level street and take your foot off the gas pedal and put the car in neutral, you'll slowly coast to a halt. We say to ourselves, "Well, sure, as soon as I stop having the engine apply a force to the car to push it along, it stops." Or if you are playing a game of pool, you can smack the cue ball and cause any of the other balls to move for a bit, but soon they either drop into a pocket or bounce aimlessly off a bumper and slow to a stop. These everyday physics experiments seem to tell us that the natural state of matter is to be at rest.

But our naive analysis fails to factor in the braking effects of frictional forces. We may attribute the slowing and stopping of moving objects to a natural tendency on their part, but in fact the friction generated by the car's tires making contact with the road, and the pool balls rolling on the tabletop, causes this action. Newton, being the genius that he was, saw through the apparent reality to the underlying truth: Unless acted on by an external force, the natural state of matter is to continue on its initial, straight-line path indefinitely.

Think of a receiver who has caught the ball on the run and is comfortably out in front of all defenders, or a running back who has broken through the line into the secondary and finds himself in the open field. Typically, in this situation, the ball carrier will make a beeline for the end zone, stopping only after he's crossed it. (And sometimes not even then. Many fans will recall one of the most memorable plays in the history of *Monday Night Football*, on November 30, 1987. Bo Jackson, who was playing for the Los Angeles Raiders and celebrating his 25th birthday, took a handoff late in the second quarter from quarterback Marc Wilson and accelerated past the entire Seattle Seahawk team for an explosive 91-yard touchdown run. He crossed the goal line and kept on running—with the ball—right up into the locker-room tunnel!)

The First Law also says that the more massive an object is, the more it wants to continue doing what it's doing and the less likely it is to be deflected, slowed down, or sped up by an outside force. One example of this mass effect in action is the touchdown scored by William "The Refrigerator" Perry in Super Bowl XX. Chicago Bears head coach Mike Ditka had sent in the Fridge as a fullback on a third-and-goal running play from the 1-yard line in the third quarter as the Bears were dismantling the overmatched New England Patriots. Starting in the backfield to work up momentum, the Fridge took the handoff from quarterback Jim McMahon and surged into the end zone. Considering that Perry, normally a defensive tackle himself, weighed in at 318 pounds that day, only a massive force applied by a very large, well-placed defender would have altered his course substantially and stopped the touchdown. Not surprisingly, that didn't happen.

Newton's First Law also provides the reasoning behind why quarterback sneak plays work if well timed. The offense uses the huge mass of its interior linemen as a battering ram against the defensive line. The surge forward can be slowed or stopped, but often not before the one or two yards necessary for the play to succeed are gained.

You may be saying to yourself, "Gee, this sounds like he's telling me that it's better to be a big fast guy than a small slow guy. Well, duh! I knew that already." Of course you did. Not everything in physics is counterintuitive. But having said this, the First Law does have some not-so-obvious consequences. Consider an experiment I do as a lecture demonstration for my first-year physics classes. Take two bricks, one made of wood, the other of lead. If they're the size of typical bricks, the lead one will weigh about 30 pounds; the wood brick, 2 pounds. Now support each brick carefully on two 8-ounce Styrofoam coffee cups set mouth down. (You'll have to be careful to avoid crushing the cups as you place the lead brick on them.) From a height of 1 yard, drop a 1-pound weight straight down on the center top of both bricks. Which brick will do a better job of crushing the coffee cups under the blow of the falling weight?

When I ask my students to predict the answer, typically more than 90 percent say that the lead brick will do more damage. That's just plain old common sense, right? But remember what

Newton tells us: The more massive an object is, the more it wants to do what it was doing before. In the case of these bricks, what they were doing before was sitting at rest before being hit. The lead brick, being about 15 times more massive than the wooden one, is much less interested in moving. As a result, it accelerates less violently into the coffee cups and does a lot less damage.

So while, generally speaking, it *is* better to be a big fast guy than a small slow guy, that isn't the whole story. Incidentally, for those of you who like to bet on football games, you may want to consider a new moneymaking opportunity. The next time you and your buddies are watching a game together, get out some Styrofoam cups and a couple of bricks . . .

THE FRIDGE'S ESSENTIAL
CHARACTERISTIC: MASS

Newton's Second Law, which is really the basis of all classical physics, says that the force applied to an object is the product of the object's mass and its acceleration. Mathematically, we say that $F = ma$. We can use this formula to appreciate just how much force is expended when one player hits another, and how much force exerted over a given time would have been needed to stop Refrigerator Perry from scoring on that running play. First, though, we must define our terms.

The mass (m) of an object is basically the amount of matter—the number of atoms—it has in it. Mass and weight are connected, but they are not the same thing. An astronaut who weighs 250 pounds on earth has a mass of 114 kilograms. On the moon, the same guy weighs only 40 pounds, but he has the same number of atoms and, hence, the same mass: 114 kilograms. While mass is a measure of how much matter an object has, weight is a measure of how strongly this mass is attracted by gravity to whatever planet or asteroid the object happens to be sitting on.

We talk about how big an offensive guard or defensive tackle is in terms of his weight, but when we are trying to determine how hard he can hit the guy across the line of scrimmage, what we actually care about is his mass. Fortunately, in the earth's gravitational field, the player's weight and mass are linked by a simple conversion formula: 2.2 pounds is the equivalent of 1 kilogram. We will routinely use the term *pound* (or pound-mass) when referring to mass, for the sake of sanity and simplicity. Just remember that a pound of mass is the mass associated with an object that weighs 1 pound on earth.

ACCELERATION, SPEED, AND POSITION: KINEMATICS

After mass (m), the next variable in our equation $F = ma$ is acceleration (a). Acceleration should not be confused with speed. Speed tells us how rapidly an object is changing its position, while acceleration tells us how rapidly this speed is changing in time. Notice that this definition of speed is no more complicated than "distance equals rate times time"; the distance you travel is equal to your average speed (or "rate") multiplied by how long you've been going. In football, we use speed to measure the time a player needs to travel a certain distance: "He runs a 4-7 40." This means he can run 40 yards in 4.7 seconds from a standing start. His average speed is thus 40 yards divided by 4.7 seconds, or 8.5 yards per second. (This is equivalent to 25.5 feet per second.) We talk about a player's "average" speed because over the course of a run he is likely to speed up or slow down.

Another term related to speed is *velocity*. Velocity is a speed specified in connection with a direction. Put another way, speed tells you how fast you're moving along the path you're taking, while velocity tells you your speed and the direction in which you're going. The units of both speed and velocity are feet per second, abbreviated "ft/s." The units of acceleration are feet per second per second. Since "feet per second" is a speed, a rate of change of speed over time must have another "per second" in it. In physics we write this as "ft/s^2," but we can also spell it out as "feet per second squared."

To put all these terms in football perspective, let's consider Jerry Rice running a deep pattern, moving from right to left. We'll use the yard markers that he passes to help us keep track of things. Starting from rest at the line of scrimmage, his own 10-yard line, he increases his speed over a period of 2 seconds until he reaches his top end—say, 32 feet per second. During the time that his speed is increasing, he runs from the 10 to about the 20. His acceleration is roughly 16 feet per second squared over this stretch. This means that after the first second, he is running with a speed of 16 feet per second. After 2 seconds he's running at his maximum speed of 32 feet per second. In other words, his speed is increasing at the rate of 16 feet per second every second. Rice's top-end speed of 32 feet per second tells us that once he's running flat-out, he'll travel 10 yards in a little less than a second. While his *speed* is now 32 feet per second, his *velocity* is 32 feet per second *to the left*.

These quantities—speed, velocity, and acceleration—are the elements that make up the branch of physics called kinematics, or the science of motion. Galileo was the first scientist to make a comprehensive study of kinematics. Newton's Second Law is important to us here because it connects motion (kinematics) with the "dynamical" quantities of force and mass.

FIGURING OUT
THE FORCE OF A BIG HIT

Remember that we entered this whole discussion about acceleration and speed to begin to appreciate, using Newton's Second Law, the force involved in a big hit from the likes of a Jack Tatum, "Mean" Joe Greene, or to use a more recent example, Ray Lewis. So let's get down to it. What is force? Simple. According to Newton's Second Law, force is the thing that speeds mass up or slows it down—in other words, gives it an acceleration or deceleration. (This is strictly true only for one-dimensional motion. In two-dimensional motion, it is possible for forces to change an object's direction but not its speed. For example, when a running back is given a sharp, impulsive blow from the side by a defensive tackle, the hit doesn't necessarily slow him down, but it will deflect his path. We'll discuss this issue more in chapter 3.)

A force must be exerted by something that is in contact with the object being accelerated—some active agent like a string, a spring, a finger, or an enraged Dick Butkus. This may sound trivially obvious, but the misunderstanding of this point is the cause of much confusion. When talking to my students in introductory physics classes, I often hear things like "The force of the truck accelerating causes the collision," or "His speed exerts a force on the opposing player." Acceleration and speed do not produce force—only physical objects in contact with the mass in question can exert a force on it. *Strings, springs, fingers, Butkus!*

Another important point: the things touching the object don't need to be visible. The earth, for example, can exert a force on you through something a bit more tenuous: its gravitational field. We can visualize this force, though, in the same mechanistic way we do strings and fingers. The earth grabs you with gravitons—elementary particles that provide the mechanism of the pull of gravity. This is the only intangible force we need consider in this book. Every other agent of force we deal with will be very tangible—as real as sack specialist Michael Strahan closing in on a panicked quarterback.

For the purpose of this book, we will specify force in pounds. In doing so we will have to modify our equation for Newton's Second Law a little bit, to $F = (\frac{1}{32})ma$. The reasons for this additional numerical factor are historical. In the old English system, mass was defined in terms of a unit called the slug. A force of 1 pound was defined as that which would give a mass of 1 slug—which weighs 32 pounds on earth—an acceleration of 1 foot per second squared.

The pound is still the preferred unit of force, but slugs aren't used much anymore; they've been replaced with pounds of mass. Thus, if we want to use pounds for the mass in Newton's Second Law, we have to stick in the factor of $\frac{1}{32}$; a force of 1 pound will give a 1-pound mass an acceleration of 32 feet per second squared. We just have to be careful to distinguish between pounds of mass and pounds of force. To minimize the confusion, we'll call a pound of mass a "pound-mass," abbreviated "lbm," and a pound of force simply a "pound-force," or "lbf."

With that bit of delightful physics trivia out of the way, we're finally ready to do our first calculation. With just how much force *did* Dick Butkus hit running backs? Let's apply Newton's Second Law and find out. The force Butkus exerts on his opponent is proportional to the ball carrier's mass times his acceleration: $F = (\frac{1}{32})ma$. Let's say that Butkus is facing off against a fullback of a similar mass—245 pounds (lbm). The back hits a hole opened up by his hardworking offensive line and he's running hard, so his initial speed is about 10 yards per second, or 30 feet per second. Then Butkus enters the picture, and the play comes to a crashing halt. The back's final speed, immediately after the hit from Butkus, is zero. The duration of the hit, from the first contact of pads to the point when the back's forward motion stops, is about two-tenths of a second. (We can determine this by analyzing a slow-motion replay of the hit.) Dividing the speed change by the time interval over which it occurred gives us the acceleration of the back—or, rather, the deceleration, as his forward motion is stopped cold: $a = (0 \text{ ft/s} - 30 \text{ ft/s})/0.2 \text{ s} = -150$ feet per second squared. (The minus sign tells us that a is actually a deceleration.) Now all we have to do is multiply by the ball carrier's mass (lbm) to find the force acting on him: $(\frac{1}{32}) \times (-150 \text{ ft/s}^2) \times (245 \text{ lbm}) = -1,150$ pounds of force, or about three-fifths of a ton in the backward (negative) direction. That's the weight of a small adult killer whale, and that's the force Butkus exerted on his hits. No wonder they call football a contact sport!

A FORCE TO BE RECKONED WITH

For perspective, let's consider how big a deceleration of -150 feet per second squared is by comparing it with the acceleration that an object of the same mass would experience falling out of a fifth-story window. In this case, the object is acted on by the force of gravity alone (we'll neglect air resistance here) and accelerates under its influence.

The interesting thing about the force of gravity is that it is proportional to an object's mass: $F_{gravity} = (\frac{1}{32})mg$. This is the direct result of Newton's Universal Law of Gravitation, which Sir Isaac developed to describe the motion of the planets. We'll use this law in a much more down-to-earth way. The more massive something is, the harder gravity pulls on it. This is why, in his heyday as a player, the Fridge weighed nearly twice as much as placekicker Martin Gramatica does; he had twice as many atoms to make up his mass. Interestingly, the g in the equation $F_{gravity} = (\frac{1}{32})mg$ stands for the acceleration due to gravity, and has a value of 32 feet per second squared, which is essentially a number constant over the surface of the earth. Any force that causes this acceleration is thus said to be a force of 1 g on the body in question.

If a 245-pound object were to fall out of a fifth-floor window, it would accelerate downward with a value of a = g = 32 feet per second squared. The neat thing is that this is true for any object with any mass—*all* falling bodies experience the same acceleration under the pull of a given gravitational field. This was exactly what Galileo suspected and is said to have demonstrated in his famous (and probably apocryphal) experiment in which he dropped a light solid sphere and a heavy solid sphere off the Leaning Tower of Pisa. Astronaut David Scott demonstrated the same principle in his feather-and-hammer drop.

In the collision between Butkus and the ball carrier, the unfortunate fullback experienced a force causing his body to have more than *five times* this acceleration—a force greater than 5 g's. A deceleration this big requires a lot of force for a given mass, and Butkus obliges.

Still, such collisions don't leave Butkus unscathed. In his quick burst through the line, the runner has built up quite a head of steam and exerts some force of his own on Butkus. Butkus's speed, also about 10 yards per second, is reduced to zero at the point of contact. Here, though, we have to assign to his initial speed a minus sign because he is moving in a direction opposite that of the ball carrier before the hit. Notice, however, that because of the change in sign of his initial velocity, his acceleration actually turns out to be positive: a = (0 ft/s − [−30ft/s])/0.2s = 150 feet per second squared. Butkus has a mass of about 245 pounds (lbm), so the Second Law tells us that the force on him is 1,150 pounds (lbf).

Notice that the collision we've been analyzing is relatively symmetric, meaning that Butkus and the ball carrier have roughly the same initial magnitude of speed, albeit in different directions. They also have the same final speed: zero. The opposite signs on the forces correspond to the differences in their respective accelerations—minus 1,150 pounds of force for the fullback, plus 1,150 pounds for Butkus—and so the magnitude of the force each one feels from the collision is the same.

A second point is one that is fairly obvious but still crucial: Butkus exerts a force only on the ball carrier, and vice versa. Neither player can exert a force on himself; neither can do anything to cause his own acceleration directly. This sounds strange, but it's true. We'll explain what we mean momentarily.

ENTER THE THIRD LAW

In the example above, both players feel the same force during the hit. After all, they're of similar size, speed, and acceleration. But what happens when a huge, fast player attempts to take out a small, stationary player? The bigger player exerts a bigger force than the smaller player, right? For the answer to this question, we must turn to Newton's Third Law. And the answer it gives is an unequivocal *no!*

Newton's Third Law says that whenever two objects collide, no matter what their individual masses, no matter how fast they're going, they always exert the same amount of force on each other, but in opposite directions. Mathematically, this can be written as $F_{12} = -F_{21}$, where we read F_{12}, for example, as "the force that body 1 exerts on body 2." The minus sign means, again, that the forces have opposite directions.

This idea may at first sound crazy to you. If Warren Sapp, running at full tilt, sacks a small guy like Doug Flutie, who is standing stock-still in the pocket, getting ready to throw, how can they possibly exert the same force on each other? After all, Flutie is the one who goes flying, right?

Well, they can and they do. You can convince yourself of this by conducting a physics experiment in the comfort of your own backyard. Take a piece of rope and connect a spring balance to each end. Now have a tug of war with a friend as each of you hangs on to one of the balances. No matter how hard you try, the readings on both scales will always be equal. If you try to pull harder and make your scale read more, your friend's scale reading will increase as well. Ease off, and the readings drop together.

MOMENTUM AND IMPULSE

How can we understand this counterintuitive idea? Perhaps the best way is to see how Newton's Third Law leads to another fundamental law of physics, the Conservation of Momentum. It is this law that is going to explain why Flutie and not Sapp goes flying.

The word *momentum* is used a lot, and usually imprecisely. In order to keep our thinking clear, we must, as before, carefully define our terms. In physics, we define momentum to be the mass times the velocity (or speed): $p = mv$. (We designate momentum with the letter p to conform to tradition and to avoid confusing it with m for mass.) Momentum is what Newton called "the quantity of motion." In any collision on the football field, the total momentum of the two players is the same before and after the hit—a principle that is extremely useful in analyzing collisions.

In addition to momentum, we want to consider impulse, which is really just the change in an object's momentum. If one object strikes another, we say that it has delivered an impulse to that body that is equal to the change in the second body's momentum as a result of the collision. The impulse is equal to the product of the time over which the collision occurs multiplied by the average force exerted on the body. Here's the punch line: Since both objects (players) exert the same force on each other but in opposite directions ($F_{12} = -F_{21}$) during the collision, and they do so over the same time interval, they must deliver equal but oppositely directed impulses to each other. This means that one player gains exactly the same momentum that the other loses, so the net change in the momentum of the two players is zero.

This idea is referred to as the principle of Conservation of Momentum, and it is one of the most important rules in physics. Put another way, it says that when any two objects interact with each other (and forces due to other external objects, such as a third player or friction from the ground, are negligible), the sum of their momenta will not change over the course of the interaction. In determining the change in momentum of an object, we must be very careful to remember the directions of the motion involved. We keep track of this with our plus and minus signs. This can get a little confusing if we're not careful. In the previous discussion we would say that technically Butkus's momentum *increases*, even though he is slowed to a stop in the collision. This is because his momentum has changed (increased) from a very negative value to zero.

Conservation of Momentum is illustrated nicely in a toy you may have seen at one time or another called "Newton's Cradle." It consists of five steel ball bearings, each about an inch in diameter, suspended in a row by nylon threads attached to supports on either side of the balls. If you raise one of the balls on the end and let it go, it hits the remaining four and kicks the ball at the opposite end up to roughly the same height as that from which the first ball was dropped. The momentum of the first ball is simply transferred down the line to the last ball.

Actually, the last time I saw a Newton's Cradle in action was not at a toy store but in the 2004 NFC playoff game between the Eagles and the Falcons. This was the fourth playoff game in a row for the Eagles, the first three having resulted in losses. The Eagles needed to score first to make a statement that the outcome of this game would be different. Fortunately, minutes into the first quarter, with the ball on Atlanta's 4 after a quick, purposeful drive by McNabb and company, Dorsey Levens carried the ball through a gaping hole opened up by right guard Jermane Mayberry and barreled down to the 1, where he and about seven Atlanta defenders collided. Unable to bring Levens down, this ungainly group milled about for a bit, scratching and clawing at each other with no visible momentum forward, backward, sideways, or down. Disgusted with the whole mess, Mayberry, who had been standing at the 5 watching this tango, decided to take things into his own hands, and ploughed into the milling crowd from behind. A chain reaction of collisions ensued, and Levens popped out of the group at the other end—and across the goal line.

Let's use Conservation of Momentum to analyze the Sapp–Flutie altercation. (See **Figure 1-1**.) Sapp gets Flutie in his sights as he blitzes into the pocket with a speed of about 24 feet per second. Flutie, happily oblivious to imminent disaster, is standing still in the pocket, looking for a receiver to get open. Flutie's velocity is zero, and he has a mass of 180 pounds (lbms). Sapp, weighing in at 310 pounds, storming past his blocker and closing quickly from the blind side, pops Flutie *hard*—so hard that Sapp falls to the ground after contact has been made, rolling on the turf at perhaps 3 feet per second. Flutie, of course, is knocked in the same direction as Sapp was traveling before the collision. How quickly is *he* moving? The initial momentum of the two

BEFORE COLLISION

AFTER COLLISION

Figure 1-1. A diagram of the Sapp–Flutie altercation. The velocities of both players are indicated before and after the hit. Momentum is conserved in the collision.

players is due entirely to Sapp and equals his mass times his velocity: 7,440 lbm*ft/s (the asterisk means "times"). Remember that total momentum is the same before and after the collision, so the momentum of Sapp and that of Flutie combined after the hit must also be 7,440 lbm*ft/s. Knowing Sapp's final velocity, we can calculate Flutie's final speed: 31 feet per second. The small guy really does go flying!

This same principle—the Conservation of Momentum—was used a few years ago by physicist John Fetkovich of Carnegie Mellon University to analyze the Immaculate Reception. Working from the classic NFL Films footage of the play, he noticed a crucial feature of the collision between Tatum and Fuqua: just before the two players and the ball converge on the same spot on the field, Tatum is sprinting upfield, opposite the direction in which the ball is moving, while Fuqua is running laterally and a bit downfield. Even more importantly, Fuqua's arm is extended toward the ultimate intersection point in order to make the catch. Fetkovich estimated that the ball's velocity was about 55 ft/s downfield, and about 30 ft/s upfield after the rebound. Given the positioning of the players and the motion of the ball, it is unlikely that it could have hit anything other than Tatum's torso or head, or Fuqua's outstretched hand or forearm. But its rebound speed—and Conservation of Momentum—tell us that Tatum must have been the one to touch it. The much lighter mass of Fuqua's extended arm and the fact that it was moving away from the ball could not have provided the impulse necessary to reverse the ball's direction. This situation is analogous to rolling golf balls at a #2 wood which is moving rapidly towards the ball in order to hit it, versus a light iron which is receding from the ball before the collision. The rebound from the wood will be much more violent.

Interestingly, Fetkovich later came across NBC footage of the same play, what *Pittsburgh Post-Gazette* reporter Byron Spice calls "football's version of the Zapruder film." In this video, the ball clearly hits Tatum. So while it's nice to have the confirming evidence, there was never a controversy as far as physicists were concerned.

WHERE COACHING COMES IN: CENTER OF MASS AND TORQUE

So far, the physics ideas presented in this chapter are as applicable to ricocheting billiard balls as they are to colliding running backs and defensive tackles. But we know there is more to a football game than inanimate masses colliding with one another. What are the fundamentals of blocking and tackling taught by coaches? And how, when they execute correctly, can small, quick defensive players sometimes lay out big running backs? Now that we have used basic physics to define the limits of what *can* happen, let's look at the practical applications of physics in tackling and blocking.

The first, most basic instruction coaches give players about tackling an opponent goes something like this: "Keep your feet apart, stay low with your head up, and drive upward and through the opposing player." In order to understand why this technique is so effective, we now take up two new physics ideas: the *center of mass* and *torque*.

Let's consider torque first. Simply put, torque is the rotational equivalent of force. In the same way that force causes a mass to accelerate along a straight line, torque causes objects to rotate about a pivot line, sometimes called the axis of rotation. The bigger the torque, the more effective it is at causing the object to which it is applied to rotate about its pivot line.

We can illustrate these concepts by considering the simple act of opening a door. The door rotates about its pivot line, defined by its hinges. Let's now apply a force of, say, 5 pounds to the door, with the direction of the force being perpendicular to the plane of the door. If the door is 3 feet wide, and we apply the force at its edge farthest away from the hinges, the "lever arm" of this force is 3 feet. The torque we apply to the door is 3 feet × 5 pounds = 15 pounds-force feet (lbf*ft). Alternatively, we can apply the same perpendicular force to the door, but now at a distance of 0.1 foot from the hinges. Now the lever arm is 0.1 foot, and the applied torque is 0.5 lbf*ft. Which torque is more effective in opening the door? It doesn't take Tom Landry to figure this one out—it's the first one. The value of the applied torque simply tells us how effective the force we're applying to the door is at opening it.

Now consider a different possibility. We'll apply 5 pounds of force to the door as before, and we'll apply it to the edge of the door farthest from the hinges, just as before. But now we change the direction of the applied force so that it points directly through the pivot line, parallel to the plane of the door. Common sense tells us that now the force we've applied isn't going to open the door at all. How does this relate to the torque? The lever arm of the force is now zero, because the force direction extends through the hinges themselves. Thus, the torque is zero; that door isn't going anywhere.

Torque by itself doesn't tell us much about tackling unless we combine it with an understanding of a player's center of mass. An object's center of mass is essentially the point through which we consider the pull of gravity on that object to act. This is why the center of mass is also referred to as the center of gravity. Most people have a basic concept of where the center of mass of an object lies—roughly at the object's center. And most people know that the admonishment "Keep your center of mass low!" means, roughly translated, "Crouch down, you!" This isn't wrong, but in order to really understand the center of mass, we need to determine the "center" of an object a little more carefully.

A player's center of mass is roughly just below his rib cage, on his vertical center line. When a player assumes a wide stance and crouches down to make a hit, his center of mass lowers (but remains in his torso area). If we constructed his helmet of lead, the center-of-mass point would move up perhaps an inch, because his head area would now be heavier. If we put the same amount of lead in his shoes (and how many of us haven't seen linemen who seemed to have this problem?), his center of mass would move down a few inches.

Here's the bottom line. When tackling or blocking, the reason to stay low and drive upward through the opposing player is so that you can control his motion by exerting far more torque on him than he does on you. Newton's Third Law still holds. You exert the same *force* on him as he does on you, but by using your knowledge of centers of mass, you can completely dominate him in terms of torque. This idea is illustrated in **Figure 1-2**.

When two linemen meet at the line of scrimmage, or when a linebacker comes up to make a tackle on a running back or short-route receiver, the two players initially exert equal magnitudes of force on each other as soon as they make contact. The force the linebacker exerts runs roughly along the line of his body and up through the ball carrier's torso. The ball carrier exerts a force equal in magnitude but opposed in direction ($F_{12} = -F_{21}$). The equal forces that they exert on each other, however, do not result in equal torques.

The ball carrier exerts a force on the linebacker that extends along the line connecting the linebacker's center of mass and his effective pivot point—the point of contact between his back foot and the ground. The defender is thus very stable under the force

F_{21}

F_{12}

PIVOT POINT

PIVOT POINT

Figure 1-2. Player on left lowers his center of mass and drives up and through the ball carrier or blocker at right. The two players' centers of gravity are indicated with solid black bursts. Pivot points occur where the players' feet contact the turf, indicated with an X.

Figure 1-3. Lateral forces are less effective at destabilizing a player whose stance is low to the ground. His feet act as pivot points for his body—like hinges on a door—and come into play depending on the direction of the force applied by the opposing player. His center of mass is indicated.

from the runner. It's as though the ball carrier were trying to "open a door" by pushing along a line that passed through the hinges. On the other hand, the defender has a large lever arm—a large amount of leverage—with the force that he exerts on the ball carrier, who rotates rapidly about his point of contact with the ground as a result of this torque, becoming unstable under the unexpected rotational motion. At the least, our linebacker will stand the ball carrier up, effectively halting his forward motion. Ideally, the ball carrier is completely bowled over and loses the ball in the process. In this kind of a hit, the coach's admonishment to "keep your head up" doesn't affect the amount of torque delivered directly, but it does help the tackler to follow through with the motion that delivers the torque. It also minimizes the risk of neck injury.

As for how far apart to keep your feet as you set up to make the tackle, a good rule of thumb is to plant them slightly wider than shoulder width. This, again, relates to stability, but now we're talking about stability in the lateral, side-to-side sense. Consider **Figure 1-3**. Anytime you fail to hit an opponent head-on, your body will experience lateral forces upon contact. If your feet were close together at this time, there would be significant leverage for these lateral forces about the point of contact between your feet and the ground. With your feet spread, however, the pivot point is whichever foot is opposite to the point of contact between the lateral force and your body. Because your body is low, below this point of contact the leverage for the lateral torque is small, and the tendency for your body to rotate off the tackle is minimized. Again, the crucial point here is that the tackler must keep his center of mass as low as possible.

The physics of blocking and tackling is the physics of executing the basics. All the fancy science we're learning won't do a team much good if players don't execute. So let us now descend to the place on the field where putting into practice all this theory is a win-or-lose proposition: the Pit.

"THIS IS A GAME OF INCHES AND SECONDS."

—SAM WYCHE

CHAPTER 2

THE PIT

The glamour and excitement of football swirls around the big plays made by the star running backs and wide receivers, but much of the real work in football, the dirty work, gets done at the line of scrimmage by the offensive and defensive linemen. These guys have a name for their place of employment: the Pit. Play in the Pit isn't pretty. Vicious head slaps, violently thrown forearms to the face, the ever-present danger of leg-snapping chop blocks, clawing, biting, and some of the most blasphemous verbal abuse you'll never hear on network TV are just a sampling of the professional hazards that guards and centers and tackles endure. Under such circumstances, it would seem that only brute force would have any effect. But the biggest brute on the field is no match for the laws of physics.

A classic example of how the fierce struggle in the Pit wins games is the 1967 championship game between the Green Bay

Packers and the Dallas Cowboys, forever known in pro-football lore as the Ice Bowl. Played in Green Bay on the last day of the year, the temperature was a balmy $-15°F$, and the biting wind made it feel like 48 below. The extreme cold made any sort of pass very risky, so the battle centered on the line of scrimmage. Conditions were made even worse by the heating system that had recently been installed to keep the grass from freezing. The morning of the game, when the tarp was lifted from the field, the water that had condensed underneath it overnight immediately froze into a solid sheet of ice, making footing unreliable at best.

The young and talented Cowboys had been challenging Green Bay in recent years for league domination, and with 5 minutes remaining in the fourth quarter, it seemed that they might finally achieve their goal. Green Bay was trailing, 17–14, as they got the ball on their own 32-yard line. A brilliant drive led by quarterback Bart Starr, however, brought the Packers to the Cowboys' 1-foot line with 16 seconds remaining, a third down, and no time-outs. Common sense dictated a pass in this situation, as an incomplete pass would stop the clock and allow time for a field goal to tie the game. But Coach Lombardi was getting cold and wanted to go home sooner rather than later. He told Starr to call a 30 Wedge. Faking a handoff to his running back and lunging headfirst into the Pit behind his surging linemen, Starr took advantage of Newton's First Law and snuck the ball into the end zone. The Packers won, 21–17, and headed off to Super Bowl II, where their relatively easy handling of Oakland in the much warmer confines of Miami's storied Orange Bowl would make them two-time world champions.

In this chapter, we'll descend into the Pit to see how physics governs the play at the line of scrimmage, and how it often determines the outcome of the face-off between the offensive and defensive armies that do battle there.

OFFENSE ALWAYS HAS THE ADVANTAGE

At the line of scrimmage, the rules of football always give a built-in advantage to the offense. The offensive players know the snap count, and they know ahead of time what it is that they want to do after the ball is hiked to the quarterback. This raises the issues of reaction time and kinematics. In order to focus on these, we will assume that neither the offensive nor the defensive linemen are penalty prone, that the offense remembers what the snap count is (this can be a problem for certain players), and that the defense isn't stunting or doing anything unsportsmanlike (heaven forbid!), like trying to draw the offense offsides.

We will start by considering a simple sequential snap count such as "Hut one! Hut two! Hut three!" If the center executes perfectly, he will begin moving the ball toward the waiting hands of his quarterback the instant he hears the beginning of the designated snap signal. He can do this because the rhythmic cadence of the quarterback's calling allows him to anticipate when the final hike signal will be called. The rest of the offensive line (and indeed the entire team), assuming they also execute perfectly, will fire off the line of scrimmage at this exact same moment.

Not knowing the snap count, the defense can only react to motion by the offense. This means that their own motion will begin sometime after the offensive line moves. The interval between when defensive players see the motion of the offensive line and when they actually begin to move is called, appropriately enough, their reaction time. The reaction time of human beings varies considerably from person to person and can also depend, in any given situation, on his or her state of alertness. In the Pit, of course, we are generally dealing with superbly trained athletes who are extremely interested in reacting as quickly as possible to the snap of the ball. As a rough estimate, we can assume the reaction time of most players to be about 0.2 seconds.

What can happen in this one-fifth of a second? Let's take some of the kinematics we learned in chapter 1 to figure out the possibilities, using statistics typical of big, fast pro linemen—somebody like the St. Louis Ram's Pro Bowl tackle Orlando Pace.

WHAT DOES IT MEAN TO "EXPLODE" OFF THE LINE?

For this analysis of the first few moments of a play from scrimmage, we need to know two important quantities: the top speed of the linemen and their maximum acceleration off the line of scrimmage. Both will have a significant bearing on how a play develops. In a short yardage situation, it can make the difference between 6 points and a field goal, or between a first down and a punt.

A good time for a big guy like Orlando Pace in full pads is 5 seconds in the 40. This is nothing for a defensive end, but it's impressive for an offensive lineman. Even though in the Pit there is usually little call for sprinting, particularly on the offensive side of the ball (setting up a screen pass is one exception), we can use this information to calculate a player's acceleration off the line if we make two reasonable assumptions based on studies of human running. First, we'll take his initial acceleration to be essentially constant; i.e., we assume that his velocity is increasing in proportion to the time since he fired off the line. (Remember how it worked with Jerry Rice back in chapter 1.) Second, we'll assume that he accelerates for the first 2 seconds of his run—we call this the *boost phase*, during which his acceleration is a_B (B for boost)—after which he'll run at constant, top-end speed (v_{max}) for the duration of the sprint. We'll call this second part of the run the *cruise phase*.

We know that the distance covered during the boost phase added to the distance covered during the cruise phase equals the total distance covered during the sprint: 40 yards. The respective times required for each segment of the sprint add up to the total time: 5 seconds. The science of kinematics tells us that, starting from a dead stop, the distance Pace travels while accelerating equals one-half the value of his acceleration times the square of the time over which the acceleration occurs. This tells us how far he runs in the boost phase. From chapter 1 we remember that distance traveled at a constant speed equals that speed times the time expended ("distance equals rate times time"). This holds only during the cruise phase, when Pace isn't accelerating, i.e., when his speed is constant. Putting all of this information together, we can now determine v_{max} and a_B. The gory details of this solution are omitted here, but the result is that Pace has a top-end speed of 30 feet (or 10 yards) per second and an acceleration of about 16 feet per second squared.

We now want to concentrate on Pace's acceleration off the line. His acceleration that we've deduced from his sprint time is about half that of a falling object (32 ft/sec^2). How much force does Pace need to exert to give himself this acceleration? Newton's Second Law, $F = (\frac{1}{32})ma$, is immediately applicable. Pace's playing weight is about 325 pounds. Thus, to give himself an acceleration of 16 feet per second squared, he needs to exert a force of (325 lbms $\times$ 16 ft/sec^2 $\times$ [$\frac{1}{32}$]) = 163 pounds. (This is where the squat-thrusts come in handy.)

Before we get back to reaction time, we should discuss this force in a bit more detail. It is correct to say that Pace accelerates off the line of scrimmage as a result of the force he exerts. But it is *not* the force that *he* exerts that causes him to accelerate. It is instead the reaction force of the ground on him. This is a direct result of Newton's Third Law. Remember that Pace can't exert a force on himself—he exerts a force on the ground with his feet as he fires off the line. According to Newton's Third Law, this means that the ground must exert an equal but oppositely directed force on him. Pace's force on the ground is *backward*, so the force of the ground on him is *forward*. It is this force that causes him to accelerate forward.

We should also note that after 2 seconds, Pace has essentially stopped accelerating. This means that the net force that the ground exerts on him during the cruise phase of his sprint is zero. During the boost phase, both feet push backward on the ground, causing a net forward force. During the cruise phase of the run, though, each foot, upon touching down on the turf, first pushes forward in a braking motion and then, as the body's center of mass passes over the foot's point of contact with the ground, switches to a backward push. The net impulse that the foot exerts while it is in contact with the ground is zero.

Having calculated all the kinematic variables in our problem, we are now in a position to appreciate the crucial importance of reaction time. Pace knows the snap count, and he fires off the line the instant the center snaps the ball. Who should happen to be waiting for him across the line? None other than Shaun Rogers! Let's say that the great Lions defensive lineman is keying on Pace's hand in contact with the ground in anticipation of Pace's initial motion off the line. Two-tenths of a second elapses between when Rogers sees Pace's hand begin to move and when he can react. During that time, Pace is accelerating toward him at a rate of 16 feet per second squared. How fast is Pace moving at the end of this boost phase? After 0.2 seconds, Pace's speed will be 0.2 seconds × 16 feet per second squared = 3.2 feet per second. In other words, before Rogers can move a muscle, Pace is heading toward him at more than a yard per second.

At this rate of acceleration, how long does it take Pace to meet up with Rogers and close the 1-yard gap of the neutral zone? For this we remember that under a constant acceleration, distance increases as the square of the time multiplied by half the acceleration. Setting distance = 1 yard (3 feet), we calculate a time of 0.61 seconds. Actually, assuming that Rogers starts to accelerate toward Pace at 0.2 seconds into the play (his reaction time), they will make first contact in a somewhat shorter time, 0.52 seconds (assuming that Rogers' acceleration equals that of Pace). At this point, Pace is moving at a speed of 8.3 feet per second. Rogers, starting 0.2 seconds after Pace, has a speed of only 5.1 feet per second. Thus, Pace has a big advantage in speed and momentum, in addition to the fact that he knows what he wants to do with his opponent tactically; Rogers can only react. When Rogers and Pace collide and stand each other up, Pace is the master of the situation.

This kind of offensive advantage is of particular importance in situations where not much yardage is needed. Remember the Ice Bowl. The ball was 1 foot from the goal line. Two-tenths of a second after the ball was snapped, the Dallas defense was faced with an oncoming Packer offensive line sporting about 5,000 lbm*ft/sec momentum and wanting nothing more than to move that ball 13 inches forward. Given that the Cowboys weren't even moving yet, this kind of momentum is hard to stop.

The offensive reaction-time advantage is retained even if the quarterback uses a staggered count. Not knowing exactly when the hike signal will be called, the offensive line will begin the play with a reaction-time delay of about one-fifth of a second after the snap signal is called. But the defense must still react to the motion of the offensive line. They can't react to the snap signal, even though they hear it at the same time as the offense, because they don't know what it is.

WHY AREN'T YOU TACKLING, DUMMY?

We turn now to another physics concept crucial to the wars in the Pit: energy. Nowhere is energy more important in football than in the collisions between the offensive and defensive behemoths who live on the line of scrimmage. This is especially true for the success of a play like the quarterback sneak that ended the Ice Bowl.

You may think you understand what energy is; a lot of guys think they're good poker players, too. As physicists, though, we want first to carefully define what we mean by energy before we begin using it in our analysis. There are actually two kinds of energy we'll be discussing: kinetic and potential. For the time being we'll concentrate on kinetic energy. As you might guess from the name, kinetic energy is the energy associated with an object in motion. The definition we will use is this:

An object has kinetic energy if, by virtue of its speed,
it has the ability to do work.

When you see the word *object* in definitions like this, just think *lineman*. An alternative but equivalent definition for kinetic energy is:

The change of an object's kinetic energy
equals the work done on it.

Okay, this seems to make sense. But wait a minute—we haven't yet defined *work*.

In its simplest form, work is *the product of a force and the distance along which the force acts*. Put mathematically, we say W = FD.

To see how this equation is applied, let's consider Orlando Pace again, except that this time he's on the practice field, hitting the tackling sled—a regimen that the denizens of the Pit are all too familiar with. If he moves the sled from the 40-yard line to the 50-yard line and applies a force of 100 pounds the whole way, our equation tells us that he has done a total of 30 feet × 100 pounds of force = 3,000 foot pounds-force of work.

This all seems simple enough, and it is, but things can get a little tricky if we don't pay careful attention. Next consider Mr. Pace exerting the same 100 pounds of force on his sled, except that now he has the rest of the Rams line standing on the sled's back deck. Pace can push till the cows come home, but that sled isn't going anywhere. After 10 minutes of sweating and straining, how much work has he done? None whatsoever! The sled hasn't moved, and so the value of D in our equation $W = FD$ is zero, and so is his amount of work. (Even though Pace hasn't done any work on the sled, he is still awfully tired. His muscles, by continually stretching and contracting, *have* done work on different parts of his body at different times, but not any that resulted in any organized mechanical motion. We'll take this up again at the end of the chapter.)

GET THE SLED OUT!

This raises another question: What if Pace works hard pushing the sled across a grassy field for 10 yards at a constant speed of, say, 3 feet per second? At the beginning and end of the 10 yards, the speed of the sled hasn't changed, so we would be hard-pressed to argue that its energy has increased. Nonetheless, Pace has certainly done work on it. This fact seems to violate our second definition. If Pace has done work on the sled, why hasn't its energy increased?

Once again, our conundrum is solved by Newton's Second Law. The sled, moving at a constant speed over those 10 yards, isn't accelerating. In other words, $a = 0$. If the sled's acceleration is zero, the force acting on it must also be zero, according to our equation $F = (\frac{1}{32})ma$. But Pace is exerting 100 pounds of force on the darn thing! Some other force must be canceling out Pace's force, but where is it coming from? The answer, of course, is the grass. The frictional force of the grass on the bottom of the sled acts in the opposite direction, and the *net* force acting on the sled is zero. It is always this net force that determines the acceleration of a given mass in the Second Law.

Now think about the work done on the sled by the grass. If Pace is exerting a force of plus 100 pounds on the sled and it isn't accelerating, the grass must be exerting a backward force of minus 100 pounds on the sled. Note, however, that the sled is moving forward, i.e., in the *positive* direction. To calculate the work that the grass does on the sled, we still use equation $W = FD$, but now the force has a minus sign, so the work that the grass does is a negative 3,000 pound-force feet. Thus, the total work done on the sled, the sum of the work done by Pace *and* the grass, is zero. Since no net work has been done, the sled's kinetic energy did not increase, as we suspected.

Before we continue, I want to clarify one point that often confuses beginning physics students. The force that the grass exerts on the sled is equal in magnitude and opposite in direction to the force Pace is exerting on the sled, but the two forces are not the force/reaction "force pair" required by Newton's Third Law. Third Law force pairs must act on different objects, and these two forces are acting on the same object—the tackling sled.

What is the Third Law twin of the force that Pace exerts on the sled? It is the force of minus 100 pounds that the sled exerts on him. Note that over the 10-yard push, the sled, through this force, does minus 3,000 pound-force feet of work on Pace. Is it just a coincidence that the grass is pushing backward on the sled with the same force that the sled is pushing backward on Pace? Yes! We constructed this example so that the sled isn't accelerating, but it would have been possible to have Pace push harder, say, with 130 pounds. In that case, the grass would still push back on the sled with the same frictional force, about 100 pounds, but the sled would have pushed back on Pace with 130 pounds. It is only the Third Law pairs that must have the same magnitude. (The reason that the force of the grass stays roughly the same is due to the physics of friction, which we'll take up in chapter 8.) In this new situation, the sled would accelerate due to the unbalanced force acting on it.

THE ENERGY OF MOTION

Let's turn the discussion now to kinetic energy itself. Kinetic energy, K, is defined in terms of the object's motion (or velocity) as $K = (\frac{1}{64})mv^2$. We can see now where the first definition of kinetic energy comes from. If we do work on something, this will speed it up, giving it energy of motion, assuming nothing else is opposing our force. This process is reversible. Once the object has kinetic energy, it can do work on something else by applying some force to whatever it hits over some distance. This is how bullets work. This is why it smarts when the coach throws the clipboard at you.

Since kinetic energy represents an object's ability to do work, the units of work and kinetic energy are the same: pounds of force times feet—lbf*ft. According to our formula for kinetic energy, this is equivalent to one lbm*(ft/s)^2.

Kinetic energy and momentum are similar in several ways but are also crucially different. If a player has momentum—that is to say, if he has mass and speed—then he has kinetic energy as well. But the two quantities are related differently to the force that produces them both. The force with which one player hits another dramatically increases the victim's kinetic energy and his momentum. But the momentum increase is the product of the tackler's average *force* over the duration of the hit times the time interval over which the hit occurs. We remember what this product is called: impulse. Multiply that same force by the *distance* over which it's applied and we get the work, W, that the tackler does on the ball carrier, and hence the carrier's change in kinetic energy. In other words, momentum and kinetic energy are both produced in a collision by force, but momentum involves the product of that force times the collision's time, while kinetic energy involves the product of that force times the distance over which the collision occurs.

As an example of how this works, let's consider tackle Lincoln Kennedy, who had a playing weight of around 335 pounds. How fast would he have to run to have the same momentum as a cannon-shot pass thrown by a quarterback like Brett Favre? In a pinch, Favre can throw a football at 100 feet per second, or more than 68 miles per hour. A football weighs about 0.91 pounds, so this translates to a kinetic energy K = 142 (lbm*ft^2)/sec^2. The ball's momentum is 91 lbm*ft/sec. Now, how fast does Kennedy have to run to have the same momentum? Crawl is more like it: about 3¼ *inches* per second! At this speed, his kinetic energy is 0.388 (lbm*ft^2)/sec^2, almost 370 times less than that of the ball. This tells us that kinetic energy is much more dependent on speed than is momentum.

There is another important difference between kinetic energy and momentum. In a collision between two players where we can neglect all forces except the ones they exert on each other (physicists call this an "isolated" collision), the sum of the two players' momenta is *always* conserved: it has the same value before and after the collision. We saw this back in chapter 1; it follows directly from Newton's Third Law. However, the sum of their kinetic energies is *rarely* conserved. This can easily be seen by considering two players of equal mass and opposite speeds that collide with each other in midair. If they wrap each other up and fall to the ground, both of their final speeds are zero. Their total initial momentum was zero too, because one of their velocities was positive while the other was negative. Their final momentum is obviously zero; momentum is conserved in their collision. The sum of their initial kinetic energies is a large number, but the sum of their final kinetic energies is zero.

The two most important conservation laws in physics are the Conservation of Momentum and the Conservation of Energy. But we just said that kinetic energy isn't conserved in collisions between two football players. Are the laws of physics suspended for football? No, they aren't.

Let's consider a collision between linebacker Brian Urlacher and running back Clinton Portis. When Urlacher smacks into Portis, a number of things happen. Their protective pads slap together and bend. Their helmets go *crack!* Their jerseys rub violently. Fingernails gouge eyeballs. All of these things cause the atoms making up the various parts of their uniforms and bodies to speed up. When you hear a *crack*, that means air molecules have been put into motion to form a sound wave. When two forearm pads scrape past each other, they generate heat through friction. If something heats up, that means its molecules are moving faster. All of these interactions amount to one basic result: the two downed players have made the atoms in and around them move faster than they were moving before. This means that the disorganized kinetic energy of these individual atoms has increased by exactly the same amount that has been lost in the "organized" kinetic energies of the two players. Kinetic energy hasn't been lost, it's just been rearranged.

THE GAME IS GETTING BIGGER

We're now ready to understand just how important kinetic energy is to what's happening in the Pit. The Rams set up with six men on the line, including the tight end. The Lions, suspecting a running play, have five down linemen head-to-head with the Rams. This is where Mr. Pace earns his money. The offensive linemen fire straight ahead off the line, with the strong-side guard and tackle opening up the hole for the running back.

How much energy does the offensive line generate during this play? We remember that our offensive guys are moving at a speed of about 8.3 feet per second as they collide with the defensive line. If we assume that they have an average mass of 290 pounds, each player makes contact with the enemy having an average kinetic energy of 313 lbm*ft²/s². Multiply this by the number of linemen, and we have a total offensive line kinetic energy of 1,875 lbm*ft²/s² at the instant they meet up with the defense. This is a considerable amount of energy to apply toward controlling the line of scrimmage.

To put all of this in perspective, let's compare this amount of energy with the kinetic energy of some other moving objects. A one-and-a-half-ton pickup truck with the same energy would be moving at about 6.3 feet per second, or the speed of a brisk walk. A bullet fired from a .357 magnum handgun has about 300 lbm*ft²/s² of kinetic energy, so we would need to unload a full revolver chamber into the defense in order to expend the same amount of energy on them. We might even suggest that the result would be no more deadly than getting hit by some of the offensive lines that Dallas put on the field in the mid-1990s.

While we're down here in the Pit, it is interesting to note that one way in which the game of football has evolved over its long history has been in the size and speed of its players, a factor that has had a real impact on strategy. **Figure 2-1** on page 62 shows how the size and speed of linemen in the NFL have increased since 1920. Since no data for the top-end speed of this group are readily available, I estimated the percentage change using world-record times in the 100-meter dash over the past century. The average mass of offensive linemen has increased by 60 percent, and their top-end speed has increased by about 12 percent, meaning that the average kinetic energy dumped into the Pit on any given play has essentially doubled since 1920.

If the kinetic energy being expended in the Pit has increased so significantly, one might expect the number of injuries suffered by players to have increased as well. There are, of course, other factors affecting injury rate. The new artificial playing surfaces have increased the number of foot, ankle, and knee injuries but haven't had a big effect on other types of major injuries. (The physics of artificial turf will be discussed in chapter 8.) In fact, the occurence rate of major injuries—concussions, broken legs, death (the kind we might expect to get from violent collisions with massive fast and strong guys)—has not increased very much since 1920. This is primarily the result of the increased kinetic energy being countered by dramatic improvements in the design and materials used in protective equipment. (In chapter 7 we'll discuss how this equipment works.) Injury rates are also lower for linemen in general because they rarely build up a head of steam prior to collisions the way, say, defensive backs do.

HORSEPOWER IN THE PIT

There is another quantity in physics that can help us to think about how energy is expended in a game of football: power. Power is defined as the rate at which energy is expended. The faster a lineman does work on another player, the more power he exerts. Power equals the work done during a given time interval divided by that time interval. We specify power in units of horsepower, which was originally defined by the rate at which a healthy horse could do work and is equivalent to the raising of 1 ton by a height of 3.3 inches in 1 second. For the sake of comparison, let's consider our one-and-a-half-ton pickup powered by a 200-horsepower engine climbing San Francisco's Lombard Street, which has a slope, or grade, of 1:2 (rise/run). (This is a very steep hill; remember the

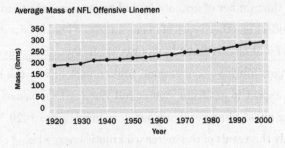

Average Mass of NFL Offensive Linemen

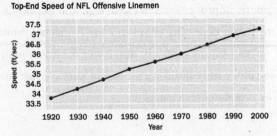

Top-End Speed of NFL Offensive Linemen

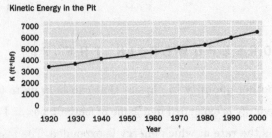

Kinetic Energy in the Pit

Figure 2-1. Graphs show how the size and speed of linemen in the NFL have increased since 1920. The average kinetic energy dumped into the Pit on any given play has essentially doubled over this time period.

chase scenes from *Bullitt*?) The engine does work against the pull of gravity at a rate of 200 horsepower (neglecting friction in the vehicle's drive), corresponding to a speed up the hill of 56 miles per hour.

The metric unit of power is the watt, named after the Scottish inventor of the steam engine, James Watt. Most electrical appliances you have sitting around the house have power ratings in watts to tell you how fast they will run up your electricity bill. You don't pay for power, though—you pay for energy. That's why your bill specifies the number of kilowatt-hours you used last month, not the number of kilowatts. A kilowatt-hour is the energy a 1,000-watt machine expends in 1 hour. There are 746 watts in a horsepower.

So what is the power of our offensive line when the ball is snapped? The six men on the line in our previous example expended 1,875 lbm*ft^2/s^2 in about one-half of a second. This corresponds to an average of 6.8 horsepower. That works out to each player putting out 1.1 horsepower—which seems like an incredible amount of power out of one guy, and it is.

It's interesting to compare this amount of power with some other values for various types of human activity. Let's consider a person of average size lounging on the couch, watching football on television. If this person has an average metabolic rate, he'll expend about one-seventh of a horsepower just to keep his heart pumping, his brain twitching, and his body temperature at a normal 98.6°F. This is a baseline power, the amount of energy your body needs to expend just to stay alive. On the other end of the spectrum are two kinds of athletic activity: sustained exertion and explosive exertion. The greatest amount of power a superbly trained athlete can expend over an extended period—two or three hours at most—is about three-fourths of a horsepower. This is the rate of power exerted by male marathon runners, or by Bryan Allen—the guy who pedaled his way into history with the first human-powered flight across the English Channel in his craft the *Gossamer Albatross*, not the pro hockey player of the same name. For extreme exertion over short periods, the human body can exceed the output of a laboring horse—if only for a few seconds. We've shown that the average power of our linemen, from hike to hit, is 1.1 horsepower, although it can be even higher.

Let's consider once again Orlando Pace firing off the line of scrimmage with an acceleration of 16 ft/s². In the first tenth of a second, his speed increases from 0 feet per second to 1.6 feet per second. Thus, his kinetic energy increases from 0 pound-mass-feet squared per second squared (lbm*ft²/s²) to 13 lbm*ft²/s². We can calculate his power over this first tenth of a second: 130 (lbm*ft²)/sec² per second, or about ¼ horsepower. This is less than the average horsepower we calculated for him earlier. What about his power in the last tenth of a second before he hits Shaun Rogers? At 0.42 seconds his speed is 6.7 feet per second; at 0.52 seconds it's 8.3 feet per second. This change corresponds to a power over that last 0.1 second interval of about 2.2 horsepower. Pace is exerting power that is changing with time. What we find in a general analysis of this type is a new physics formula: $P = Fv$, or power equals force times velocity. (To do this analysis properly we would need to use the methods of calculus, yet another of Newton's inventions.)

When Pace runs his 40-yard dash, he exerts a constant force against the ground of 163 pounds as he accelerates, so that the ground exerts the same force against him. His velocity is increasing proportionately with time during the boost phase, so his power is too. When he reaches his maximum speed of 32 feet per second, the power he exerted during the last tenth of a second of the boost phase was 9.1 horsepower! The average power he exerts during the boost phase is his total change in kinetic energy divided by the duration of the boost phase: 4.7 horsepower.

Over the course of a typical football play lasting 5 seconds, Pace will expend roughly the same amount of average power that he would during the boost phase of a sprint. This means that his body does 4.9 horsepower × 5 seconds = 24.5 horsepower-seconds of work. If he could sustain his power output of 4.9 horsepower, he could move our pickup up Lombard Street, by himself, at a speed of more than 1 mile per hour. Working together, all six linemen could move it up that same hill at about 8 miles per hour, the speed of a fast jog.

In a typical football game, there are, on average, about 100 plays. How much energy do both teams expend over the course of four quarters? Let's take that figure of 23.5 horsepower-seconds we calculated and work up a rough estimate. Over the course of the play, our six men on the line will expend about six times this amount of energy: almost 150 horsepower-seconds. The remaining five guys on the team, the so-called skill players—a moniker that used to infuriate us linemen—probably expend at least 15 horsepower-seconds each over the course of the down. (This leaves open the question of why, if they are working so much harder, the linemen get paid so much less, but I digress . . .) Thus, we have a total energy of roughly 225 horsepower-seconds being expended by the offense on the play. The defense works just as hard as the offense, so we'll double this number to get the total energy expended per down: 450 horsepower-seconds, or roughly 130 times the energy the offensive line expends in the first half-second after the ball is snapped. To get the entire energy expenditure for the game, just multiply by 100. This gives us almost 45,000 horsepower-seconds.

Now let's play a different game with our pickup truck. If we expended 45,000 horsepower-seconds to lift it up into the air, how high could we go? We can figure this out because we know its weight: 1.5 tons. One horsepower-second will lift 1 ton 3.3 inches in the air. Thus, 45,000 horsepower-seconds will lift our truck 100,000 inches in the air, or one and a half miles. That's a lot of energy!

LASAGNA, THE BREAKFAST OF CHAMPIONS

Where does all the energy and power expended during a football game come from? From food, of course. To answer the question from a physics point of view, we now have to consider the idea of potential energy. All of the kinetic energy that is expended on the football field comes from potential energy—the energy contained in the food that players eat. Potential energy is basically stored energy waiting to be released in the form of kinetic energy—the ability to do work. Perhaps the simplest type of potential energy is gravitational potential energy.

Imagine a football sitting on the ground. It isn't moving, so it doesn't have any kinetic energy and can't do any work. If you pick it up and hold it 2 yards off the ground, it still isn't moving, so it has no kinetic energy. But it does possess *stored* kinetic energy by virtue of its height. Why? If you drop it, it will fall to the ground. Just before it hits, it will have plenty of kinetic energy. By doing positive work on the ball—by lifting it—energy is stored in the form of *gravitational potential*. Gravitational potential energy does not play a big role in football, but another kind of potential energy does. It's the kind our offensive line uses to move defenders out of the way: the chemical potential energy stored in food. It's easy to figure out how much potential energy is stored in food—just count the calories. To do this we have to know two things. First, a physics calorie is not a food calorie. One food calorie equals 1,000 physics calories. Second, a physics calorie is a unit of energy just like the horsepower-second or foot-pound force. The calorie unit is used more in problems relating to heat and thermodynamics. A physics calorie is defined as the amount of heat required to raise the temperature 1 degree centigrade in 1 gram of water. (A gram of water takes up a cubic centimeter, or 1 milliliter, of volume.) It is equal to 0.0056 horsepower-seconds.

Taking a brief detour, we should note that heat *is* energy. The physicist James Prescott Joule showed that one could heat water in an insulated container by stirring it with a paddle. The paddle was driven by a pulley mechanism driven by a falling mass. The potential energy of the falling mass was converted not into the kinetic energy of the mass but into the heat energy of the water. The hotter something is, the more violent the motion of its atoms or molecules. As we have seen, when two players collide, a lot of the organized kinetic energy associated with the speeds of their centers of mass is turned into the disorganized kinetic energy of heat caused by friction.

We can now calculate how many (food) calories the average person must eat just to sit on a couch all day and watch football. His expenditure of power is about ⅐ horsepower. A day is 86,400 seconds long, so he will burn about 12,000 horsepower-seconds per day. This corresponds to a daily energy intake of 2,200,000 physics calories, or 2,200 food calories. This calculation is really just a statement of the conservation of energy in disguise; if a person burns more potential energy than he takes in, he will start burning potential energy in the form of fat—he'll lose weight.

Now let's calculate how many calories Orlando Pace burns during the course of a game. Assuming he plays every play on offense, he'll be in about 50 plays. Using our previous estimate of 23.5 horsepower-seconds per lineman per play, we multiply by 50 to get the total energy expended in one game: about 1,200 horsepower-seconds. This works out to about 220 food calories. Now that doesn't sound like much, given all the energy expended and power produced over the course of the game, and indeed this number is deceptively low. This is the food energy that went directly into useful mechanical work done during each play; it doesn't take into account the fact that extreme physical exertion has the effect of kicking up the metabolism—the rate at which the body burns energy—long after the immediate exertion is over.

On game day, guys the size of Orlando Pace or Shaun Rogers can easily consume more than 10,000 food calories without gaining an ounce. This is the number of calories in three glasses of orange juice, half a pound of hash browns, five eggs over easy, a big ribeye breakfast steak, two Snickers bars for a morning snack, a couple of helpings of lasagna, two bowls of coleslaw, six buttered dinner rolls, two quarts of Gatorade, three apples, three pork chops, two baked potatoes with cheese sauce, three bananas, a big ice-cream sundae, two slices of peach pie, and a small dish of low-fat yogurt sprinkled with artificial sweetener for a midnight snack. (In the case of my own football career, the diet involved a lot of beer and Twinkies as well.)

In the old days, players were fed protein-rich foods such as lean steak before games. Team nutritionists now realize that the ideal meal before games is something heavy in carbohydrates, like lasagna, because this kind of food is most readily turned into blood glucose, the body's purest form of fuel. Don't be discouraged when your personal trainer tells you that a heavy workout burns off only 50 calories per hour. She's right, but that doesn't take into account that your body will be kicked into high gear, metabolically speaking. You'll actually end up burning a lot more calories well beyond that hour of mechanical work.

There's one other reason that 220 calories is a low estimate for the total energy Pace burns during the game. When he and Rogers collide and stand each other up, their respective kinetic energies drop to almost zero as they do their tango in the Pit, grunting and clawing the whole time. Technically, they aren't doing any mechanical work on each other, because neither is exerting a force through any significant distance. Nonetheless, common sense tells us that if they kept this up for several minutes, they'd both have "worked pretty hard," to use an imprecise phrase. In other words, they'd both be sweaty and tired. In fact, their muscles *have* expended energy; it's just not energy that resulted in observable mechanical motion. Instead, work was done due to small parts of the muscle contracting and relaxing—expending force over short distances—within the muscle tissue. This factor, coupled with the huge metabolism linemen have as they cycle through short bursts of physical exertion, is what ramps their caloric requirement up to such high levels.

In point of fact, players who make their living in the Pit can lose as much as 15 pounds over the course of a game—more than 5 percent of their body mass. This weight loss is not the result of burning fat, however, but due to water loss—sweating. They can easily gain that back after the game by guzzling lots of liquids.

"HE ALWAYS EMPHASIZES NOT THROWING INTERCEPTIONS."

—QUARTERBACK KEN ANDERSON ON BILL WALSH

CHAPTER 3

THE WEST COAST OFFENSE EXPLAINED

In modern football, the single most important strategic element is the passing game. College quarterbacks who can scramble, run the ball like an I-back, and, through sheer determination and leadership, inspire an otherwise mediocre team to victory after victory won't ever play quarterback in the NFL unless they can throw the football. As we have seen, successful running plays require strength, agility, and toughness. The physics issues in running plays focus on energy, force, and power. Passing plays add another dimension to the game: the open space of the field. The details of how players move in two dimensions are now crucial.

One of the greatest quarterback–receiver combinations in the history of football was Joe Montana to Jerry Rice. Perfecting what is now referred to as the West Coast offense, Montana and Rice terrorized defenses around the league in the 1980s and 1990s with a relentless passing attack. In this chapter we're going to analyze, from a physics perspective, what made them such a deadly offensive threat. We will also consider other elements that are crucial to the open-field game: evading a defender, chasing a ball carrier, getting open, rolling over a defender, and throwing on the run.

TIMING IS EVERYTHING

The West Coast offense was developed by a series of visionary coaches starting in the 1960s, the most notable being Sid Gillman of the San Diego Chargers, John Rauch and Al Davis of the Oakland Raiders, and Bill Walsh of the San Francisco 49ers. The central idea is to substantially replace the running game with short passes, intermingled with a few longer passes when the tactical situation warrants. There are typically a large number of eligible receivers downfield on each play, including running backs and one or two tight ends as well as the wide receivers. The quarterback is the linchpin in this offensive scheme. He must read the defense accurately at all stages of the developing play and pass the ball to his most opportune receiver.

One of the most effective quarterbacks in the game today, Peyton Manning, is a master of West-Coast-offense style passing. In fact, he recently broke Dan Marino's single season touchdown passing record, which had stood for 29 years. With five, and sometimes even six eligible receivers downfield (including two tight ends and his running back, Edgerrin James), Manning's strong right arm can send a bullet pass to virtually any part of the field.

A crucial element of these passes is timing. In order to maximize the chance of an open receiver gaining a lot of yards after he catches the pass, he must not have to alter his speed or route in order to catch the ball. This means that the quarterback must throw the ball to a predetermined point in space, with his selected receiver arriving at that point at a predetermined time. Ideally, he must know where all of his receivers and their defensive counterparts will be before and after the snap, and make a lightning-quick decision about which receiver to go to. This is one of the elements that makes the West Coast offense so difficult to learn and endless practice so crucial to its success. Indeed, some of the playbooks for teams that employ the West Coast offense bear a striking resemblance to some of my graduate physics texts in both girth and complexity. Attempts to disrupt this split-second timing are the reason teams playing against the Colts tend to have a lot of illegal contact and defensive holding penalties called on them.

The exacting nature of the routes run by the receivers and the crucial importance of timing make kinematics the main physics issue for these plays. Let us consider the classic 44-yard pass that Montana threw to Rice in Super Bowl XXIII, which turned out to be the longest reception in Rice's record-breaking day of 11 catches for 215 yards. It's a classic because it illustrates Rice's incredible ability to get open and the accuracy with which medium- to long-range passes must be thrown in the West Coast scheme.

The line of scrimmage is at Cincinnati's 18-yard line, with the ball on the right-side hash mark. Rice lines up wide right, 12 yards from the sideline. Montana takes the snap and rolls right. The first point to mention here is that every movement in this roll-right is choreographed; Montana has obviously practiced this motion repeatedly so that each step, each arm motion, is intentional. The second thing to notice is that Rice, using his incredible agility and quickness, has already gotten past several Cincinnati defenders and is sprinting away from the line of scrimmage essentially unimpeded.

QUICKNESS VS. SPEED

Now is a good time to break away for a discussion of the difference between quickness and speed. Jerry Rice, at his peak, could run the 40 in about 4.6 seconds. That's not an incredible time—it's fairly typical for a wide receiver. What separates Rice from the pack of his peers is his ability to get open, and that has to do with quickness, only one aspect of that 40-yard time. If we use the same 2-second boost-phase model for Rice that we did with Anthony Pace in the previous chapter, we get an acceleration of 16.8 feet per second squared and a maximum speed of 33 feet (11 yards) per second. Throughout this book, we will use the word *quickness* to mean acceleration, which is the ability to change one's direction or speed rapidly. *Speed*, on the other hand, means just what it did when we defined it in chapter 1: it's the rate of change per second of a player's position along a straight line. (We're about to modify and generalize this definition to include two-

Wide receiver Jerry Rice, Joe Montana's favorite target and a key component in the 49ers' masterful execution of the West Coast offense, reaches for one of eleven receptions in San Francisco's championship win over Cincinnati in Super Bowl XXIII.

dimensional motion as well.) Roughly speaking, quickness is what is required to outmaneuver other players, while speed is required to outrun them.

NFL coaches have a deep if not necessarily mathematical understanding of these kinematic issues. There has been extensive discussion in the league over the years about how useful the 40-time is as a measure of a wide receiver's or a cornerback's worth. Players with phenomenal 40-times often get cut after summer camp, and there are quite a few guys in the Hall of Fame with mediocre stats in this department. In a one-on-one face-off between a wideout and a cornerback, it may happen that the receiver with a lot of good moves and quickness will beat his man at the line of scrimmage. But if the defensive back has a higher top-end speed, after chasing the receiver for 20 yards (called catch-up speed), he will ultimately prevail. While combines and agents tend to put a premium on a player's 40-speed, it is the top-end speed that usually excites defensive coordinators.

These differences illustrate the limits of usefulness for physics in analyzing the game of football. We can see easily from Newton's Second Law what is required for quickness—a high muscle-to-mass ratio. This will guarantee a big ratio of force to mass, or a large acceleration. The factors that make a given player capable of great speed are much more difficult to analyze and involve a very difficult problem in both physiology and kinesiology. Physics can only "constrain" the problem, i.e., tell us what a maximum possible value of speed might be, given the basic characteristics of how the human body is put together and how it converts food energy to mechanical energy. We can see, though, why it is possible for a player to be quick without being speedy. Bob Hayes of the Dallas Cowboys was one example of a guy with both.

My own career as a college football player illustrates this paradoxical situation. I was big, but I also had strong leg muscles. This meant that I could fire off the line with a good deal of acceleration and hit the defensive lineman across from me quickly and hard. This, unfortunately, was where my usefulness to the Fighting Beavers ended. Having next to no athletic talent, I couldn't run fast and I couldn't tackle.

MONTANA TO RICE FOR 44

Now let's get back to that pass play. Rice tangles with cornerback Lewis Billups off the line, and then runs a simple streak pattern downfield. Montana quickly reads his receiver options and selects Rice, knowing that failure to execute the pass accurately will, at best, cause Rice to alter his route and/or his fluid motion to re-adjust to the position of the ball. This could cause him to (a) not make the catch or (b) make it out of bounds. An even more unpleasant possibility is (c) that it could be picked off.

Just how accurate, kinematically speaking, must Montana's pass be? In other words, what is the margin for error on both the timing of the pass and the speed of the ball? For the play to work as designed, Montana must put the ball right at Rice's position the instant he is ready to haul it in. Our basic kinematic equation, distance = speed × time, will tell us all we need to know, but we'll have to be careful with timing and position in order to determine the necessary kinematic quantities correctly. To make this easier, we'll keep our own clock running to time the play—one separate from the official game clock. Our clock starts the instant the ball is snapped.

Montana takes the snap (at t = 0 seconds) at Cincinnati's 18. The ball was on the right hash mark at the beginning of play, so he starts 70.8 feet from the sideline. He takes the ball and rolls right until he's at the 11, 10 yards from the sideline. This well-choreographed procedure takes exactly 4 seconds. What has Rice been doing in the meantime? Mixing it up with his defender on the line of scrimmage and sprinting for the end zone. After getting around Billups, he accelerates steadily for the first 2 seconds of his run and maintains a steady 30 feet per second after that. This play calls for him to catch the ball 40 yards from the line of scrimmage.

Our first job is to figure out how long it takes him to get to this point. This in turn requires that we know the distance along the straight line from his starting point to the point of reception. The procedure for calculating this distance involves, among other things, the Pythagorean theorem and a Cartesian grid. For the delight of scholarly coaches, the curious, and the masochistic, I have included the details of this calculation, and a diagram of the play, in the Appendix (see page 259). For everyone else, the answer to the problem is 41 yards. Remember that Rice starts 12 yards from the sideline on the Bengal's 18. He catches the ball 5 yards from the sideline. This means that the distance he has to run is not the simple sum of his distances sideways and up and down the field; it's somewhat less than that. (This represents a practical example of the mathematical idea that the shortest distance between two points is a straight line.)

Using the kinematics we learned in chapter 1, we can calculate the time required for Rice to run this distance: 7.1 seconds, including the 2 seconds he spends dancing with the Bengals' cornerback. Since Montana is going to throw the ball at $t = 4$ seconds, this means that the travel time of the ball must be precisely 3.1 seconds. But wait! The distance the ball has to travel is the distance from Montana to the point of reception. This turns out to be 142 feet. We can't just use the straight-line distance along the ground that the ball travels though; we must take into account the fact that the ball is traveling in an arc. We will discuss this issue at length in the next chapter; for now, I'll just give you the result. Neglecting air resistance, and assuming that the ball is launched at a typical angle of 35 degrees above the horizontal, Montana has to throw it with a speed of 86.5 feet per second, or 59.0 miles per hour.

It is perhaps obvious but nonetheless important to note that Montana hasn't whipped out a calculator to work out the relevant kinematic equations while he's in the pocket. This might be possible if linebackers Joe Kelly and Leon White would take a break while he tries to remember whether it's d-squared or d-cubed in the Pythagorean theorem, but they don't oblige. No, the timing and precision are the result of an incredible amount of drill. Montana and Rice are both legendary in the NFL for their extreme work ethics. Success in football follows the famous admonition about getting to Carnegie Hall: practice, practice, practice.

THREADING THE NEEDLE

We've calculated when and at what speed Montana needs to throw the ball to Rice. Let's now take up a different problem. How sloppy can Montana get before Rice fails to make the catch? The quarterback is the key player in the West Coast offensive scheme, and he has three objectives: throw the ball at just the right speed, throw it at just the right time, and aim it perfectly. This latter goal becomes much more difficult if he's throwing on the run; we'll worry about that in the next section of this chapter. We will also assume for the sake of simplicity that Rice runs *his* route perfectly.

What is the margin for error in the speed and timing of the pass? Let's assume first that Montana throws at the perfect instant with perfect aim, and determine the tolerable error in the ball speed. To do this we will also need to designate a "catch zone" around Rice. He won't make the catch if the ball is outside this area when he gets to the flag. The catch zone is a circle roughly 5 feet in diameter, centered on the receiver. This dimension is based primarily on the length of Rice's arms, but it is also affected by the fact that if he has to turn or break stride too abruptly, his chances of catching the ball are reduced. Given these considerations, a 2.5-foot radius is reasonable.

Montana must now ensure that the speed of the ball delivers it to the catch zone, again assuming that he has thrown it at precisely the right time. We've already shown that the distance from Montana to the point of reception is 142 feet. The catch-zone diameter now creates a range of distances that the ball can travel in the allotted 3.1 seconds. If the pass overshoots Rice by the maximum allowable 2.5 feet, it will have traveled a distance of 144.5 feet in 3.1 seconds; if it is short by the largest allowable amount, it will have traveled 139.5 feet in the same time. Taking into account once again the arc that the ball has to travel, we find that the ball must be launched with an error of only plus-or-minus 0.53 miles per hour in the throwing speed of 59 miles per hour. We can now see clearly why quick-draw timing is the key to this wide-open style of play.

Assuming that Montana's throwing speed is perfect for this play (31.2 miles per hour), just how accurate does his timing need to be? Now we have to concentrate on Rice's speed, not the ball's speed. Rice's top speed during this play is 30 feet per second. He thus travels through the 5-foot catch zone in 0.17 seconds. Basically, if Montana is too slow or too fast by much more than a fifth of a second, we're talking third and long instead of 6 points.

These rather stringent limits that we've just calculated assume that neither Montana nor Rice makes any adjustments to what he is doing as the play evolves. In reality, they do see each other, and Rice will turn and watch the ball a few yards before he reaches the predesignated point of reception. This allows him to alter either his route or his running speed, so we can relax these estimates a bit. But passing in the West Coast offense does require incredible accuracy, and every team using this system strives for ultimate timing and throwing control at the level that our kinematic equations require.

SPEED IN TWO DIMENSIONS: VELOCITY

If you've had trouble slogging through some of the math we've been doing so far, you're in good company. One of the greatest physicists who ever lived, Michael Faraday, was almost completely ignorant of fancy mathematics. Like all good physicists, though, he didn't let ignorance stop him. When he was formulating his laws of electricity and magnetism, he replaced complex mathematical equations with *vectors*. That's just what we're going to do here, except that instead of worrying about volts and amps like Faraday did, we're going to use vectors to analyze a play that is legendary in the annals of football. Known simply as "The Catch," it sent Joe Montana and his San Francisco team to their first Super Bowl and has been categorized by some as the six most important yards in 49ers history.

Earlier in this chapter we discussed how, in the open field, our analysis had to account for motion in two dimensions. We began to do this by considering straight-line distances between two points with differing yardage values as well as different transverse positions on the field. In doing this, we essentially reduced the problem to one straight-line dimension again, even though we were laying it out in two-dimensional space. Now the time has come to bite the bullet and see how to really handle things in two dimensions. To do this, we will use vectors.

Vectors are arrows, nothing more. They have two critical attributes: length and direction. Up until now we've talked about a player's speed as a measure of how fast he's moving in one dimension, and we've used pluses and minuses to indicate whether he's moving forward or backward along that direction. We have referred to the combination of speeds with a plus or minus sign as velocities. We will now use vectors to replace velocities, but we'll still call them velocity vectors.

The direction of the velocity-vector arrow tells us which way the player is running, and the length tells us how fast he's moving. This means that the length of a velocity vector is given in feet per second, not in feet. Vectors are used in two or three dimensions whenever we need an arrow to provide complete information about a physical quantity. (For more on vectors, see the Appendix.)

WHAT'S "THE CATCH"?

We're now ready to use vectors to analyze Joe Montana's spectacular game-winning pass to Dwight Clark for The Catch. Vectors give us the simplest and most obvious method for analyzing a difficult problem: How do you determine where a pass will go when you have to throw it on the run? San Francisco was playing at home against Dallas in the 1981 NFC Championship. It was the fourth quarter, and the 49ers were down 27–21 with just 51 seconds remaining. Montana was in his element—he had just led the team on an 89-yard drive down to the Cowboys' 6-yard line. But then, on third down, not only the game but Montana's reputation, the team's season, and the hopes and expectations of all the 49er fans in the house were riding on the next play.

Montana takes the snap and rolls out to his right, loping along at about 12 feet per second. Dwight Clark, initially lined up as a wide-right receiver, drifts left and into the end zone (see **Figure 3-1**). The problem for Montana is that he's moving to his right ($\overrightarrow{A}$), but Clark, his only viable receiver, is downfield and to his left. This means that he has to throw backward off the wrong foot—not a well-rehearsed or simple situation. Montana needs to mentally and instinctively factor in the effect of his velocity, and he subconsciously uses vectors to do it.

While Montana must adjust his throw on the run and in a fraction of a second, we can stop and break things down on paper. To determine the direction in which the ball will go, we add Montana's velocity vector ($\overrightarrow{A}$) to the velocity vector the ball would have for the exact throwing motion he uses, but assuming that he is standing still instead of throwing on the run ($\overrightarrow{B}$). The resulting actual ball velocity combining Montana's ground speed with his throwing motion is the vector whose tail is at the tail of Montana's velocity vector, and whose tip is placed at the tip of the ball's "stationary throw" velocity vector. Adding the two vectors, we can see that their sum $(\vec{C})$ is pointing to where Clark will be in the time it takes the pass to reach him.

Montana lets fly, and the rest, as they say, is history. The 6-foot-4 Clark extends his body to the fullest, leaps into the air over a Dallas defender (who shall remain nameless), and plucks the pass out of the air. Of course, Montana does all this with the instincts of the great quarterback that he is, but ultimately he must obey the law of vector addition—and credit Dwight Clark with an incredible catch. *The* Catch.

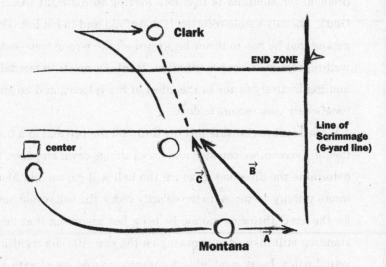

Figure 3-1. The Catch, diagrammed. Joe Montana takes the snap under center and rolls right. To get the ball to Dwight Clark, Montana must throw off his right foot moving away from his target. His velocity is represented by vector $\vec{A}$. To throw the ball the proper direction to be caught (vector $\vec{C}$), he must throw with a motion that, were he standing still, would give the ball the velocity of vector $\vec{B}$.

SURELY YOU'RE JUKING

Up until now in this chapter we've talked blithely about pass receivers running their geometrically perfect routes smoothly and effortlessly as though there were no other players on the field. Unfortunately, the defense usually isn't willing to be this accommodating.

One thing pass defenders do, of course, is try to slow down a receiver just off the line of scrimmage in order to interfere with the timing of his route. Offensive planning generally tries to take one or two hits into account, but getting hung up with a defender can seriously jeopardize the chances for a successful pass completion. When the receiver reaches the open field and wishes to avoid a downfield defender, he must take evasive action. This, of course, is particularly true once the receiver catches the ball, as the yards gained after a catch, popularly known as YACs, are essential to the success of a West Coast offense. We will consider general evasion and chase strategies in the next section, but consider now one tactical method of evasion: juking.

Juking, simply put, is a rapid change of course direction, possibly involving a change in speed, possibly repeated several times in quick succession. Of course, both receivers and running backs can juke; indeed, outstanding juking ability is a much-valued characteristic. It also leads to serious ankle and knee injuries, and, using the method of vector addition, we're about to see why.

As practical football physicists, we know what we're dealing with when confronted with changes of speed and direction in short amounts of time: big accelerations. Perhaps no one has ever juked better than Barry Sanders of the Detroit Lions (although others might bestow that honor on Walter Payton). Consider the common scenario of Sanders squirting through a hole his offensive line has opened and squaring off with a waiting linebacker. Sanders's velocity vector through the line is roughly straight ahead, with a magnitude (length) of 18 feet per second ($\vec{v_1}$) (see **Figure 3-2**). Sanders plants his right foot hard just as a head-on collision with the defender seems to be inevitable, and, literally in the blink of an eye, he is now moving at 18 feet per second *at right angles* to his initial velocity ($\vec{v_2}$). The hapless linebacker's reaction to this juke move is typical of defensive players who encountered Sanders in the open field: he crumples like a pile of wilted sweat socks.

Using vectors and the Pythagorean theorem (see the Appendix on page 259), we can show that the acceleration vector ($\vec{a}$) related to $\vec{v_2}$ and $\vec{v_1}$ has a magnitude of 127 feet per second squared. Using Newton's Second Law, we can also calculate the force Sanders has to exert on the ground to produce an acceleration of this magnitude: 880 pounds. Since all of this force is essentially acting through his right knee and ankle as he makes the cut, we begin to appreciate where ankle and knee injuries come from. Notice that this amount of force gives Sanders an acceleration of about 4 g's (4×32 ft/s^2). If he could continue accelerating at this rate for 10 seconds, he'd be moving faster than the speed of sound!

Running backs also take advantage of the reaction-time kinematics we discussed at the beginning of chapter 2. The defender will always be about a fifth of a second behind the evasive moves the ball carrier uses to get free, and this limits his ability to track those moves. This effect is enhanced if the runner can create a diversion that the defender keys on, such as looking in one direction

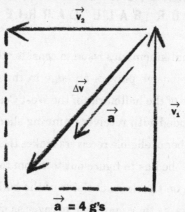

Figure 3-2. Barry Sanders, above, jukes the socks off a hapless defender. Below, velocity vectors before ($\vec{v}_1$) and after ($\vec{v}_2$) his move, connected by the change in his velocity ($\Delta\vec{v}$), yield his acceleration ($\vec{a}$): 4 g's!

or moving a forearm one way while actually moving his center of mass in another. The defender may react quickly—but to the wrong signal.

Another way to beat a defender, of course, is to simply run over him, trusting that your mass will obey Newton's First Law and insist on doing what it had been doing just prior to contact: moving in a straight line toward the goal. Common sense tells us that big runners such as Jerome Bettis, Ricky Williams, and Mike Alstott have a distinct advantage once they break through the line into the open field, and Newton's First Law explains why. A famous example of this "technique" was when John Riggins ran over Don McNeal for the go-ahead touchdown in Super Bowl XVII.

EVASION STRATEGIES FOR BALL CARRIERS

There's one final kinematics issue to consider that gives both offensive and defensive players an edge in the open field: chase strategies. One of the hallmarks of the West Coast offense is that the field is flooded with receivers running short passing routes. The large number of eligible receivers makes the quarterback's job more difficult—he has to figure out to whom to pass. But it's especially tough on the defensive side of the ball, for cornerbacks and linebackers, as they may have to cover as many as five or six eligible receivers on a single play.

Let's consider the chase strategies possible for a fake-sweep-right screen pass, where the quarterback pitches to his fullback, who then stops in his tracks and throws to a tight end 10 yards past the line of scrimmage at midfield (see **Figure 3-3a** on page 90). It's up to a free safety, even farther downfield to the tight end's right, to stop the play, because the rest of the defense has vectored in on the fullback going the other way!

To begin with, we'll make the simplifying assumption that the tight end and the free safety run equally fast. Later on we'll consider the more realistic but complex situation where the defender can run somewhat faster than the ball carrier. (This always seems to be the case in physics—the more realistic one's assumptions are, the more difficult the problem is to analyze.)

How does the cornerback maximize his chances of catching the ball carrier in the open field? We consider first the most obvious strategy: he runs straight at the ball carrier, adjusting his path so that he is always headed straight toward his man. A simple computer program can be used to solve the kinematics for two-dimensional motion with constantly changing direction. Using the output from this analysis, we immediately see the folly of this method (remember both players run equally fast). **Figure 3-3a** shows the paths of the two players for a typical evasive path of the ball carrier. As long as he runs away from his pursuer, he can always avoid being tackled. This turns out to be the case in general, even if the chase began at the same yard marker on the field. If your plan is to run somebody down by aiming straight at him, you'd better make sure you're a faster runner than he is!

Here's a better option for making the tackle. As soon as the de-

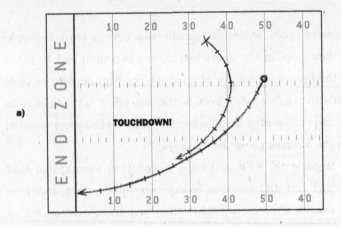

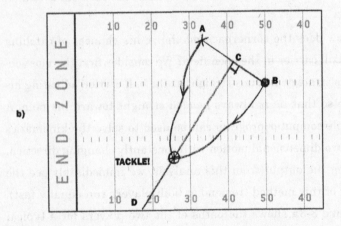

Figure 3-3. Open-field chase and escape strategies. a) Trajectories when the defender (X) runs straight at the ball carrier (•) and each has the same speed; equal time intervals are indicated by tick marks. b) Mirror chase strategy (see text). c) The ball carrier should not run away from the line CD. d) The defender should not rush the line CD. e) Two defenders reduce the maximum possible gain on the play. In c through e, arrows show the ball carrier's best-bet trajectories for gaining the most yards.

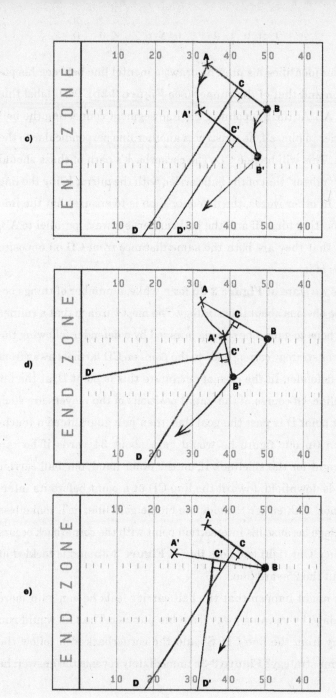

fender identifies his prey, he draws a mental line between his position and that of the runner (see **Figure 3-3b**). We'll label this line AB. At the midpoint of the line segment connecting the ball carrier to himself, he constructs another line perpendicular to the first. This will be line CD. The cornerback's path of chase should now "mirror" that of the ball carrier, with the mirror being the line CD. In other words, the defender's job is to ensure that the line connecting himself and the ball carrier is always parallel to AB, and that they are both the same distance from CD on opposite sides of it.

As we stare at **Figure 3-3b** for a while, a number of things become obvious about this strategy. The most yardage that a runner can hope to gain if he's being pursued by a defender following the mirror strategy corresponds to the point on CD farthest away from the defender. In the case of our figure this is point D, at the far sideline. Of course, if the initial positions of the players are such that point D is past the goal line, then he's guaranteed a touchdown (in our figure he would gain about 34 yards if he ran straight for the sideline). If, on the other hand, our ball carrier heads downfield, toward the line CD at a point before its intersection with either the sideline or the goal line, he'll make less yardage because his intersection point with the cornerback occurs earlier. Our tight end does this in **Figure 3-3b** and is tackled at about the 25-yard line.

It might happen that the ball carrier feels he can gain more yardage by running away from his pursuer. Thus he would run *away* from the line CD. Should the cornerback still follow the mirror strategy? **Figure 3-3c** immediately reveals the answer: he should continue to mimic his prey's distance from AB along CD, while moving parallel to AB by the same amount as the ball carrier in the ball carrier's direction. This has the effect of increasing the advantage of the defender. He now has a new set of baselines, A'B' and C'D', that yield even less maximum yardage for the offense. We can also see from this type of graphical analysis why it is a mistake for the defender to head too quickly toward CD (**Figure 3-3d**). (Aiming directly at the runner—the situation we considered earlier in **Figure 3-3a**—is an example of this.) This has the effect of producing a new coordinate system whose intersection D' with the sideline is even farther toward the goal—if it didn't intersect it already! The moral of all this for the ball carrier is: Run in a straight line to where the CD axis goes out-of-bounds.

The lesson for the defense: Follow the mirror strategy unless your quarry is increasing his distance from C along AB. If he is, decrease yours by the same amount. In the latter case, keep in mind that the baselines are constantly being redrawn in your favor. The advantage of the basic mirror strategy is that the defender doesn't have to anticipate the moves of the ball carrier. Its disadvantage is that if the ball carrier is smart, he can gain the full yardage corresponding to CD intersecting the sideline, which may be more than the defense wants him to have.

These ideas can be expanded to include the more commonly occurring situation in which there is more than one defender. We continue to assume that all of the players' running speeds are equal and that the runner still isn't lucky enough to pick up any blocks. If there are two defenders chasing the runner, they both set up their own coordinate systems and follow the mirror strategy accordingly (**Figure 3-3e**). In this case, it is in the runner's best interest to head straight to the intersection point between the sideline and that CD or C'D' axis that intersects it *farthest* away from the goal line—in our example, point D'. This is the best he can hope to do.

CIRCULAR REASONING

Finally, we take up the more difficult problem in which ball carrier and defender have different running speeds (see **Figure 3-4**). We will call the running speed of the defender v_D and that of the offender v_O. In our first analysis, we concluded that the best strategy for the ball carrier is to run in a straight line to the point where CD intersects a boundary line of the playing field. The job of the cornerback is to mirror that trajectory and intersect CD at the same point that the runner does. Thus, both men are running in straight lines at constant speed to the point at which the tackle occurs.

When the two players have different speeds, this general strategy is still the same, but the geometry changes a bit. Let's assume that the ball carrier runs in a straight line at constant speed v_O in any direction he chooses (even backward!). One can show that if the defender runs at a constant speed (v_D) in a straight line to intercept the runner, their point of intersection will lie on a circle of radius $R = v_O v_D L/(v_D^2 - v_O^2)$, where L is the initial straight-line distance between the two players. The center of this circle lies on the straight line containing A and B, a distance $D = v_O L/(v_D - v_O)$ from the initial position of the ball carrier. This means that the circle passes between the ball carrier and cornerback in their starting positions. The position on the circle at which the intersection takes place depends solely on which direction the ball carrier runs. The defender is guaranteed to intercept him simply by heading for the point of intersection between the circle and the ball carrier's straight-line trajectory.

The circles shown in **Figure 3-4** correspond to the two cases where the defender is (50 percent) faster than the ball carrier (lower circle), and when the defender is only two-thirds as fast as the ball carrier (upper circle). Depending on which case we have, it is clear where the ball carrier wants to head. If he's slower than the cornerback, he wants to get to the farthest down-field position of the "fast defense kill zone." This is represented by vector 2. Notice that this yardage gain is less than if he and the defender had the same speed, and a mirror chase strategy (vector 3) resulted in a tackle at the sideline. If he can outrun the defender, he's better off choosing vector 1, which results in a touchdown, just missing the region where he can be tackled by the slower corner. If the speed of the defender is a lot more than that of the ball carrier, the radius of the lower circle is smaller and its arc may not even intersect the sideline, let alone the goal line.

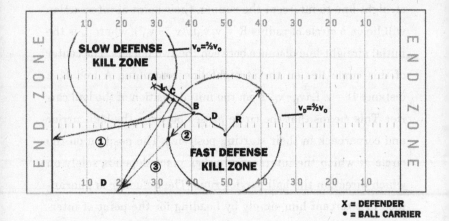

Figure 3-4. Chase strategy when one player runs faster than the other. The upper circle corresponds to the possible tackle points when the defensive back has two-thirds the speed of the ball carrier—the offense scores a touchdown (vector 1). The lower circle corresponds to the opposite situation, in which the maximum possible yardage gain is indicated by the short arrow (vector 2). In the case of equal velocities, the maximum yardage (vector 3) is that corresponding to a mirror chase strategy as in Figure 3-3b.

With $v_D = \frac{2}{3}v_O$, the ball carrier can head off in any direction he chooses, and it is guaranteed that the cornerback can catch him, although possibly not inbounds. This is because the possible interception points form a circle around the ball carrier's initial position. With the speeds reversed, there is only a limited range of directions the ball carrier can take to guarantee that the two players will meet. Basically, he has to run toward the defender to ensure the tackle. Assuming that he is smart enough not to do this, he can always avoid being tackled by simply running in a direction that does not cross this circle.

Of course, it is impossible for players in the heat of battle to make the necessary geometric constructions and mathematical calculations to follow these trajectories exactly. Skilled ball carriers (and defenders, for that matter) know almost instinctively which paths to follow in these situations. It is good strategy, though, for coaches to present these ideas to their players when they discuss with them the science of open-field running and how it can improve the productiveness of the West Coast—or any—offense.

"THE MAN WHO COMPLAINS
ABOUT THE WAY THE BALL
BOUNCES IS LIKELY THE
ONE WHO DROPPED IT."

—LOU HOLTZ

CHAPTER 4

THE FOOTBALL IN FLIGHT

The Buffalo Bills had come up short in three straight Super Bowls, beginning with a heartbreaking loss to the New York Giants in Super Bowl XXV, when kicker Scott Norwood's 47-yard field-goal attempt sailed wide right with just seconds left. On January 30, 1994, the Bills had a chance to redeem themselves and bring some relief to their suffering fans. Making an unprecedented fourth straight appearance in the NFL championship game, Buffalo faced a rematch with the Dallas Cowboys, the very same team that had handed them their lunch the previous year in Super Bowl XXVII, 52–17. The Bills were out for revenge.

Things did not go well at first for Buffalo. Dallas scored on its opening possession with a 41-yard field goal by kicker Eddie Murray. Already in the hole just 2 minutes into the game, Buffalo was running true to form. But then—a glimmer of hope! The Bills got the ball and quickly drove downfield under the direction of quarterback Jim Kelly and on the legs of Bills running back Thurman Thomas. Unfortunately, a stiffening Cowboys defense stopped the drive at the Dallas 37. Desperate to get on the scoreboard, Buffalo coach Marv Levy sent in Steve Christie to attempt something that had never before been accomplished in the Super Bowl: a field goal of more than 50 yards. (The previous record had been 48 yards, set in Super Bowl IV by Hall of Fame kicker Jan Stenerud and matched by Denver's Rich Karlis in a losing effort in Super Bowl XXI.) With a quick soccer-style kick, Christie blasted the ball through the uprights for a new Super Bowl record: 54 yards. The Buffalo Bills tied the game at 3, but they were on their way to another, somewhat more dubious Super Bowl record—four straight losses.

In this chapter we'll find out about the basic physics of a football's flight, and how it constrains what quarterbacks and kickers can do in a given situation. We'll also learn how external variables such as wind and altitude can affect a game's strategy. Once we establish these basics of projectile motion, the next chapter goes into more detail on the kicking game. In chapter 6 we'll consider the finer points of passing.

FROM THE BRUTAL
TO THE BEAUTIFUL

The physics of football is evident in all aspects of the game, from the brutal to the beautiful. In our discussions of blocking and tackling, we dwelt on the brutal; now, the beautiful: the flight of a ball. There are few things in sport more impressive than a long bomb, perfectly thrown and picked out of the air by a wide receiver working in heavy traffic. Think of that signature spiral pass coming toward the camera, the virtual trademark of Steve Sabol and NFL Films. As football fans, we want to know whether the pass resulted in a touchdown or an interception. As physicists watching the ball, we want to satisfy our curiosity about a number of nagging questions: What factors affect the distance a football can be thrown? What is the best launch angle and speed for a field-goal attempt? Why does the ball sometimes wobble on its trajectory, and does this have a negative influence on its range and accuracy? These issues go to the heart of football physics.

The way a game is going will dictate what a kicker or passer does in a given situation. So will factors such as wind and rain. If an offensive coordinator is faced with a fourth and goal on the 16-yard line, his quarterback is 1–18 that day with four interceptions, his star running back has been averaging −1.5 yards per carry, and the wind is at their backs, then he will almost certainly go for a field goal. Given that his offensive line has been as porous as Swiss cheese, his concerns about the field-goal kicker will probably not center on the guy's maximum range but, rather, whether he can get the ball off quickly in the general direction of the goalposts. A punter usually wants to maximize his range, but not always. If he's at midfield, he'll want a good hang time on the ball and sufficient accuracy to drop the ball right into what's called the coffin corner. For a quarterback directing a West Coast offense, particularly during the closing moments of the half or the game, a quick release, an ample range, and a tight spiral are crucial.

PUNTS AND PARABOLAS

When a football is thrown, kicked, or punted, it moves through the air in an arched path, or trajectory. Some of the oldest questions in the history of science relate to the trajectory of a cannonball or missile. Galileo's systematic experiments and Newton's analysis of bodies acting under the influence of gravity provided our earliest understanding of this kind of motion. The good news here is that the mental sweat we expended learning kinematics and dynamics in the first three chapters now has yet another payoff: projectile motion is just kinematics in two dimensions under the influence of gravity.

In the context of football, projectile motion defines the amount of time the ball spends in the air—its *hang time*—and the distance it travels over the ground, its *range*. Generally speaking, punters want to maximize both their range and hang time, while placekickers just want to maximize their range. A related concern of punters, kickers, and passers is accuracy, which we'll get into later.

Let's consider both dimensions of motion separately for a moment. When the ball is launched, either by hand or by foot, it moves vertically upward with some (positive) speed, slows down, stops, and then accelerates downward. Due to the effect of air drag, it hits the ground with a vertical speed a bit less than the upward vertical speed it started with. In the meantime, the ball moves horizontally along the ground at a speed that is continually being reduced by air drag. We will take up the detailed effects of air drag momentarily.

In a typical punt, the football leaves the punter's foot at an angle of 45 degrees and a speed of 100 feet per second. Plotting the ball's path, we discover that it takes the shape of a parabola, modified somewhat by the air drag forces. We encounter parabolas all the time. A parabolic shape is ideal for focusing electromagnetic waves such as light or television signals, so most satellite dishes are parabolic. The signal from outer space bounces off the surface to a single focus, where a sensitive detector picks it up. The fact that any signal that hits any part of the dish is focused to the same point greatly increases its strength—and brings football games into our homes and sports bars on Sunday afternoons and Monday nights. The same principle applies to the microphone reflectors used by TV crews on the sidelines. Likewise, parabolic mirrors are used in astronomical telescopes to focus light rays coming from distant objects into a lens, while flashlights and searchlights use the idea in reverse: they take a point source of light—the lightbulb— and direct its rays outward in a parallel beam.

Let's now analyze our typical punt, considering first the numbers that we get from the oversimplified situation of a punt in a vacuum, which would experience no air drag. From **Figure 4-1** on page 104 it is apparent that the kicked ball travels a distance of about 104 yards, reaching a maximum altitude of a little over 26 yards. What is not obvious from our graph is its hang time; this turns out to be about 4.4 seconds. All of these quantities are determined from the kinematics we discussed in the first two chapters. Basically, the hang time is the same as the time a ball thrown straight up in the air, with the same initial vertical velocity, would take to return to the height from which it was thrown. The punt's range, or distance, is this time multiplied by the ball's horizontal velocity.

Now comes the crucial idea for punters and kickers: the range and hang time depend on both the speed and the launch angle of the ball. Consider the two extremes: if the ball is launched at an

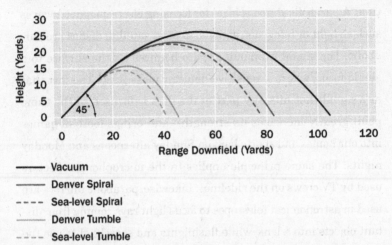

Figure 4-1. Flight of football with no air drag (in a vacuum) and with low and high altitude air drag for a tight spiral and tumbling end-over-end. The launch speed is 100 feet per second at an angle of 45 degrees. Air drag is computed for average temperature and pressure at sea level (Giants Stadium, for instance) and at an elevation of a mile (Denver).

angle of 0 degrees (horizontally) or 90 degrees (straight up), its range will be zero. The 0-degree kick with no elevation hits the ground instantly, so it can't go anywhere—a lot of onside kicks are essentially like this. The 90-degree kick goes straight up and comes right back down; it doesn't go anywhere, either. What this means, of course, is that maximum range for a given launch velocity is attained somewhere between 0 and 90 degrees. Not surprisingly, the best launch angle in order to maximize range is precisely in the middle: 45 degrees. This means that our example punt went the longest possible distance for a ball with a launch speed of 100 feet per second; the range falls off as the launch angle becomes steeper or more shallow than 45 degrees. For a 30- or 60-degree punt, the range is reduced to about 95 yards; for 15 or 75 degrees, it drops to 52 yards.

How does the varying launch angle affect hang time? For a 100-foot-per-second launch, the longest possible hang time, 6.2 seconds, occurs at a launch angle of 90 degrees. As the angle becomes shallower, this time falls slowly at first and then more rapidly. A 30-degree launch angle yields a hang time of just over 3 seconds.

Launch speed is the other important parameter. Kinematics tells us that for a given launch angle, range is proportional to the square of launch speed. Thus, as launch speed is decreased from 100 to 75 feet per second (roughly the speed difference between a pro and a high school punter), the range on a 45-degree punt drops from 104 to 59 yards. For a squib kick that leaves the foot at only 35 feet per second, the range is only about 13 yards.

Finally, figuring out how hang time is affected by launch speed is easy: halve the launch speed and you halve your hang time. A ball launched at 45 degrees at 100 feet per second has a hang time of 4.4 seconds. Launch it at the same angle at 50 feet per second and it stays in the air for just 2.2 seconds.

QUALITY HANG TIME

What are the implications of all this for tactical issues on the field? With the warning that our numbers will change a bit when we account for the effects of air resistance, we can use the results of these idealized calculations to see the general picture. Let's consider passing first. As we saw in the previous chapter, spatial accuracy in passing is critical. In other words, the quarterback must ensure that the ball's range is perfectly matched to the point of reception. He can do this by adjusting both the launch speed and the launch angle of his throw. Almost always, he will want to use a high throwing speed and the smallest possible flight time so the play develops as quickly as possible. This dictates the use of launch angles below 45 degrees that correspond to the correct range for a given throwing speed. The exception to this will occur if his receiver is particularly slow, or if there is tight coverage downfield and a good possibility that the pass might be intercepted if it falls even a little short of the receiver's outstretched arms. In this case, the quarterback will want to "put the ball up"—launch it at an angle greater than 45 degrees—either to give his receiver enough time to get where he's supposed to be, or to keep the ball well above defensive players in between him and his target.

Punters generally want the range to be as large as possible, which means kicking at an angle of 45 degrees. Things get more interesting if we're at midfield, which is generally too far out to try a field goal. On the other hand, most high school and college punters, and virtually any pro, can easily put the ball in the end zone from this far out. Thus, our punter will want to back off on range a bit and put the ball in the coffin corner, to the outside of either set of hash marks and as close to the goal line as possible. At the same time, he'll want to maximize the hang time to give his team the best chance to get downfield and cover the ball. This means elevating the launch angle above 45 degrees. Getting a good elevation on the ball is also desirable from the standpoint of keeping it above the outstretched arms of blockers on both field-goal and punting plays.

An exception to these general rules comes when an onside kickoff is desirable. In this case, the launch angle should be minimized. We'll revisit some of these issues again in the next chapter, when we have a more comprehensive model of how the ball is kicked.

WHAT A DRAG

There's a story that the guys over on the ag campus at the University of Nebraska love to tell about physicists, to illustrate how we lack connection to the real world. There was this farmer who needed to design loading ramps to get his chickens as efficiently as possible from the barn to the trucks that would carry them to market. He called up the university and asked to talk to a professor of agricultural science, but they were all busy. The secretary asked him if he would like to talk to somebody over in the physics department instead. With some hesitation, the farmer said yes. He was connected with a theoretical atomic physicist who was eager to help and promised to look into the problem. He came out to the farm, looked the chicken coop over carefully, made lots of measurements with a meterstick, and drove off, muttering to himself. After several weeks, the farmer became impatient because he hadn't heard anything. He phoned the professor.

"Just give me a few more weeks," said the physicist. "This problem is a lot tougher than I thought it would be."

The farmer called back several times over the next month and was always assured that a solution was close at hand. Finally, when he had despaired of ever getting any help, the farmer got a call from the professor.

"I have it all figured out! It was a really tough problem, and I couldn't solve it until I made a few simplifications, but I think I have it now."

"What simplifications did you make?" asked the farmer.

"First," said the atomic physicist, "assume a spherical chicken . . . "

After you stop laughing and pick yourself up off the floor, I would like to make several points in regard to this funny little story. Yes, real life is more complicated than the abstract version physicists often analyze in terms of simplified mathematical models. It is true that physicists tend to simplify problems a lot. But in our defense we usually do it because, if we didn't, we wouldn't be able to analyze or predict anything in a rigorous, mathematical way. The trick is knowing when you've gone so far that you can no longer learn anything useful about the problem you initially wanted to solve.

In the case of projectile motion, we have gotten the basic idea by resorting to simple kinematics, but reality is more complicated. Up until now we have ignored the effect of air drag on the ball, but air drag has a significant effect on the trajectory of actual passes, kicks, and punts, so we're going to have to grit our teeth and take it into account.

The basic idea of air drag isn't too complicated. Think about holding a football, dropping it, and measuring the time it takes to fall to the ground. Now let it fall the same distance through a vat of molasses instead of through the air. What happens to the time? It increases. The molasses acts to retard the motion of the football, causing it to take much longer to hit bottom than it did in the air.

Because the downward force of gravity is always opposed by the upward drag force, the total vertical force on the ball is reduced, and the acceleration due to gravity is diminished, so the ball doesn't fall as quickly. The drag force has two crucial aspects. First, it always acts in opposition to the velocity of the ball; it can never act to speed it up. Second, the drag force gets bigger as the velocity gets bigger. The faster something moves, the more adamantly the drag force opposes its motion.

What causes the drag force? It's simply the result of the molecules that make up the molasses bumping into the ball. The faster the ball moves, the harder its collisions with the molecules. It's sort of like the colliding linemen discussed in chapter 1. Imagine now that we heat the molasses up and it gets thinner and thinner. What will happen? The drag force will diminish, but it won't disappear entirely. We can think of air as being very thin molasses. Indeed, physicists think about air being a fluid in the same way they think of molasses as a fluid. It's just that air is a very thin fluid. This means that air, like molasses, can cause a drag force to act on a football. (The branch of physics relating to phenomena like air drag is called fluid mechanics.)

Before we turn back to football, let's consider another example of drag from the sport of skydiving. Air drag on the open parachute is what slows the parachutist down so that she will be safe when she lands. But what about her fall before the chute opens? The instant that the skydiver jumps out of the plane, her downward speed is zero, and no air drag opposes her motion—there's no motion to oppose! Now she begins to accelerate downward under the force of gravity. As she picks up speed, the air drag become bigger and bigger. After falling some distance, she is moving so fast that the drag force on her acting upward is equal in magnitude to the force of gravity (her weight) downward. Thus, the net force acting on her is zero, and Newton's Second Law tells us that her acceleration must now be zero as well. Since her acceleration is zero, her speed must be constant (but not zero!). This speed is called her *terminal velocity*. For humans decked out in skydiving gear with a closed chute, it's about 120 miles per hour.

When the parachute opens, the skydiver's *coefficient of drag* gets a lot bigger. This is just a fancy way of saying that for a given speed, the air can oppose her motion with a much larger force. The terminal velocity, the speed at which gravitational and drag forces just cancel, is reduced significantly, to the point that she can make a safe landing. This, of course, is the whole point of the parachute.

DETERMINING DRAG FORCE

We now want to apply these ideas to the flight of a football. It turns out that the drag force on a football is proportional to the cross-sectional area that the ball presents to the onrushing air (A), the density of the air (D), and the square of its speed (v^2). The actual equation for determining drag force is $F_{drag} = \frac{1}{2}CADv^2$, where the factor C, the "coefficient of drag," is essentially a measure of how effective the air is at opposing the ball's motion. A soccer ball, a discus, and a football all have different values of C because of their different shapes. The value of C also depends on how the object is oriented as it moves through the air and can depend slightly on the ball's speed, but this dependence is weak in the velocity ranges of relevance to football.

Note that the drag force does not depend on the ball's mass.

Now let's calculate the drag force on a football, considering all of these factors one at a time. The first factor is the ball's area (A) perpendicular to the ball's flight path. To keep things simple, we'll limit our discussion at first to the two most common situations encountered in passing and kicking: a tight spiral (most often encountered in passes) and end-over-end tumbling (usually seen in kickoffs and field-goal attempts). In the case of the spiral pass, the area we want is that associated with the diameter of the fattest part of the ball, at its middle. For a Wilson NFL regulation football, this dimension is close to 6.8 inches, so A is 0.25 square feet (the area of a circle equals its radius squared times the constant $\pi = 3.14$). End-over-end tumbling motion is a bit more involved. In this case, the ball is rotating about an axis roughly parallel to the ground and perpendicular to its flight path. This means that A is constantly changing—from the 0.25 square feet we just calculated to the value when the ball's long axis is straight up and down. This larger area turns out to be 0.41 square feet.

The next quantity to consider is the air density, D. This is the number of pounds-mass per cubic foot the air has. At sea level, D is about 0.075 pounds of mass per cubic foot. This corresponds to about 700,000,000,000,000,000,000,000—that's seven hundred billion trillion—molecules and atoms of air (mostly nitrogen, argon, and oxygen). Thus, if the ball is traveling through the air at 100 feet per second in a tight spiral, it has to push about 20,000,000,000,000,000,000,000,000—twenty *trillion* trillion!—air molecules out of the way per second. Is it any wonder that air exerts a drag force?

Finally, we consider the drag coefficient C. As mentioned earlier,

C depends both on an object's shape and its orientation in flight. It doesn't depend on size; that's taken care of by the area A. (For purists, I should note that for football velocities above 10 miles per hour, or 15 feet per second, the flow of the air around the ball is turbulent, so C is essentially independent of velocity and air density.) The value of C for a football has been measured by at least two research groups. Unfortunately, they each get results that are significantly different. We will deftly sidestep controversy here and simply adopt the more recent and extensively published results. We need to consider C for two orientations of the ball: when the long axis is parallel and when it's perpendicular to the direction of motion. In these cases, according to Rae and Streit, C is roughly 0.14 and 0.85, respectively. (They did not actually measure the latter number, but I have estimated it based on the behavior of the rest of their data.) For the end-over-end tumbling motion of the ball, we'll just use the average of the product of A and C for the two orientations of the ball when its long axis is parallel and perpendicular to the direction of motion: 0.19 feet squared. For a tight spiral pass or kick, we'll use AC = 0.035 feet squared.

FACTORING IN ALTITUDE

Our newfound knowledge of air drag lets us now see how different stadiums and environmental conditions can affect the flight of the ball—altitude, barometric pressure, temperature, and wind conditions all become factors. Players and coaches need to consider these effects when planning their game strategy. What plans, for example, do the Houston Texans need to make to play in an arid environment such as one that they would encounter in Arizona? Or how should they adjust to windy environments such as New Jersey's Meadowlands? How should their game plan change if they are traveling to Green Bay in December? And what about playing in the thin air of Denver?

As we move from the Meadowlands to Mile High, the air gets thinner, i.e., less dense. This is due simply to the gravitational pull of the earth on the air molecules. Given their druthers, the molecules that make up the air would just sit around on the grass, doing nothing, hanging out. But they have thermal energy, which gives them speed, so they jump up and down a bit. The tension between gravity's pull toward the earth and thermal kinetic energy keeps about 90 percent of our air molecules within the first 10

The "thin" mountain air surrounding INVESCO Field at Mile High, in Denver, is a boon to punters and placekickers alike.

Giants Stadium, located near sea level in the New Jersey Meadowlands, is home field to both the New York Giants and the New York Jets—and known for its shifty winds.

miles or so of the earth's surface. This all leads to a decrease in the atmosphere's density as altitude increases. (There isn't a sharp cutoff of the atmosphere at a specific altitude; a few molecules can get pretty high.) This layer of air, of course, is what keeps us alive; any atmosphere the moon ever had was quickly lost due to its weak gravitational pull.

The speeds of the individual air molecules are not all the same, but their average speed is determined by the local air temperature. The warmer the air is, the faster its molecules move. In any given batch of air, some of the molecules move at very high speed, and others barely move at all. An average speed for an air molecule is about 1,500 feet per second. In moving from the Meadowlands—or Baltimore, or Washington, or Jacksonville, or Miami, or Seattle, or San Diego, or any of the other stadiums and cities near sea level— to what's currently called INVESCO Field at Mile High (located near where the old Mile High Stadium, John Elway's house, used to stand and which is, in fact, about a mile above sea level), the air density, on average, decreases by about 20 percent. At a height of 20 miles, the air density is only 1 percent of what it is at sea level.

In order to get a quick feel for how big this drag effect can be, let's calculate the magnitude of the drag force at sea level (i.e., standard atmospheric pressure) when the ball is moving at a speed of 60 miles per hour, or 88 feet per second, in a tight spiral. Plugging in all the numbers, we get a force of about 0.31 pounds. That may not sound like much, but remember that the ball itself weighs just 0.9 pounds. In our initial calculations, only the ball's weight mattered; this new drag force can actually be comparable to (and sometimes even bigger than) the ball's weight, so we should now expect the ball to travel on a significantly different path.

SPIRALS VERSUS DUCKS

We're now ready to put all of this together by adding the weight of the ball and the drag force acting on it. The drag force will, in general, act both horizontally and vertically. It will always exert a "backward" force, opposing the ball's speed in the horizontal direction, and, as the ball ascends from a pass, punt, or kick, drag will combine with gravity to reduce its vertical speed. When the ball is descending, drag will push upward. In all cases, the drag force acts to slow the ball down.

Using Newton's Second Law and the basic kinematics equations from chapter 1, we can write a computer program to calculate the effects of air drag on spiral passes and tumbling kicks. For a visual explanation of the essential results of this analysis, let's look again at **Figure 4-1** on page 104, where we have simulated kicks launched at 100 feet per second with a launch angle of 45 degrees. As we have seen, if drag forces were zero, this launch angle would give us maximum range for a given launch speed. But until we establish colonies in space and build stadiums to keep us from being homesick for our favorite game, football will not be played in a vacuum, so let's consider four earthly situations in which air drag and altitude are factored in.

We take the typical air density of both a sea-level stadium such as Giants Stadium in the Meadowlands—any stadium below 500 feet is essentially at sea level—and Denver's stadium at Mile High. Given these two densities, the drag force on the ball has been calculated for a perfect spiral, in which the nose of the ball is locked on to the ball's trajectory curve, and for tumbling end-over-end motion. The effect of air drag at sea level is rather large. In the case of the tumbling trajectory, the range of the kick is reduced by almost two-thirds from the no-drag value. While the spiral kick is still roughly parabolic in shape, the tumbling kick has a clear asymmetry to its trajectory because it suffers much greater drag. Of course, this is also the reason why quarterbacks want to throw spirals instead of ducks: greater speed, greater range, greater accuracy.

There are two other major effects of drag that are not apparent from the figure. In the no-drag case, the hang time of the kick is 4.41 seconds; for the spiral thrown with air drag at sea level, this is reduced to 4.06 seconds. For tumbling at sea level, it is decreased by more than a second to 3.25 seconds. The kinetic energy of the kick is also bled off significantly due to the drag's frictional effect. In the no-drag case, the ball slows as it gains height, but the potential energy thus acquired gets converted back into kinetic energy on the way down. The ball ends its flight traveling with the same speed at which it began: 100 feet per second.

In the real world, the energy of the football is partially converted into the kinetic energy of the air molecules it bumps into. Imagine a semitrailer barreling into Qualcomm Stadium in San Diego (known locally as The Murph, after sportswriter Jack Murphy), which is filled to the 31st row with Ping-Pong balls. The truck will lose kinetic energy as it collides with them. The mechanism isn't hugely efficient at slowing the truck down, but it does work! In the case of our kick, the spiral trajectory has a final speed at ground level of 77.1 feet per second, and the tumbling kick ends up with a speed of only 48.2 feet per second. We can convert these speeds back into kinetic energy by remembering that $K = (\frac{1}{64})mv^2$, i.e., it's proportional to speed squared. Thus, the spiral ball has $(77.1)^2/(100)^2 = 59$ percent of its initial kinetic energy left, while the tumbling ball retains only 23 percent of its kinetic energy.

Now, like good experimental physicists, let's twiddle a knob and see what happens. It's important to twiddle only one knob at a time, so that any changes we observe can be due to only one thing. In this case, we'll twiddle the "density of air" knob by moving from Giants Stadium (elevation 60 feet) to Denver (elevation 5,183 feet). In moving from New York—excuse me, New Jersey—to Denver, the range of the spiral kick is increased by 4 yards, from 78 to 82 yards. The altitude effect is more pronounced on a percentage basis with the tumbling kick, increasing its range from 41 to 45 yards. The hang time of the kicks is not affected appreciably, increasing by 0.06 seconds and 0.12 seconds for the spiral and tumbling kicks, respectively.

What happens when a team from California or Florida that depends heavily on the West Coast offense comes into Denver to play the Broncos? How badly is their timing thrown off by the change in altitude? We can conclude from this discussion that it will not appreciably affect their pass patterns. Even if the spiral trajectory were the result of a Hail Mary pass instead of a kick, the timing change due to drag differences would be at most 0.06 seconds off—not a serious problem.

HANG TIME AND RANGE

What is the general effect of air drag on hang times and ranges? In addition to reducing the range, drag acts to lower the launch angle at which the maximum distance occurs. In a vacuum, this angle is 45 degrees. For a spiral pass on earth, however, the best angle is about 43 degrees. With end-over-end kicks it shifts even lower, to about 38 degrees. This is consistent with the videos I have taken or watched of kickers told to go for distance; they generally seem to boot the ball at about a 40-degree angle. Reduction in range for spiral passes is, as seen in **Figure 4-1**, about 20 yards for the maximum-range situation close to 45 degrees. This foreshortening is reduced to about 10 yards for launch angles of 20 degrees and 70 degrees. The Denver–Meadowlands difference of about 4 yards for spirals at launch angles around 45 degrees is reduced to only 1 yard at 20 degrees and 75 degrees. Similar reductions are seen for tumbling kicks.

Hang time for tumbling kicks is reduced by the presence of drag, with the largest deviations occurring at the highest launch angles. For example, at a launch angle of 75 degrees, the vacuum value of hang time for a 100-foot-per-second launch is 6 seconds. In Denver, this is reduced to 4.5 seconds; at sea level, it is 4.3 seconds. These differences decrease smoothly as the angle of launch gets shallower.

As for how altitude affects range, it's best to look at kickoffs as opposed to field goals or punts because the field goal doesn't have a set launch point, and a punt's recorded distance can depend a lot on how far (and which way) it rolled after hitting the ground if it wasn't caught.

We'll look at the kickoff distances in Denver and compare them with the kickoff distances in the home stadiums of all the teams Denver played in the 2001 and 2002 seasons. Using our computer model and taking into account the effect of air drag on a tumbling kick, we expect that kickoff distances should be roughly 5 yards longer in Denver, on average, than they will be in stadiums close to sea level. We need to be a little careful about this. In a given stadium, the home team will kick the ball half of the time on average. This means that if Denver has a really good kicker and San Diego has a really bad one, looking at *all* the kickoffs will bias the data in the direction we expect it to go, but for the wrong reason. We'll correct for this by using only data from kicks made by players visiting Denver, and then comparing these data with kicks they made on their home turf. Reviewing data from a large number of teams also helps to minimize the "kicker quality" factor.

In this analysis I have removed distances for onside kicks, kicks that went out of bounds, and kicks following penalties. Kicks that sail into the end zone and result in touchbacks are listed by the record keepers as having gone 70 yards. Not knowing their actual range, we have assigned them a distance of 80 yards. **Figure 4-2a** shows the distribution of home-team kickoff distances for the eight roughly-at-sea level home stadiums of the teams that played in Denver in 2001 or 2002: Miami, Seattle, San Diego, Oakland, Washington, New York Giants, Baltimore, and New England. **Figure 4-2b**, on the other hand, shows the equivalent distribution for these teams when they kicked in Denver. It is clear from this comparison that the distances in Denver are skewed well toward the end zone. It is interesting to note that the difference in the kickoff distance averages between the two distributions is 7.3 yards, plus or minus about a yard. This is consistent with the roughly 5-yard difference we predicted with our air-drag model in **Figure 4-1**, especially given that those distances are closer to 40 yards.

These results back up qualitatively what pros like punter Louie Aguiar have told me about kicking in Denver: guys who have trouble getting it in the end zone at home watch in amazement as their kicks sail past the goalpost in Denver. I have learned to regard the sincere testimonials of elite football players regarding their athletic experiences with severe skepticism, but Mr. Aguiar seems to be correct in this regard.

We can get a somewhat more detailed view of the situation by plotting kickoff distance averages versus altitude in the cities/stadiums listed in **Table 4-1** (see page 123). We have also plotted the percentage of kicks that make it into or past the end zone in the same cities/stadiums (**Figure 4-3b** on page 122). To indicate the approximate uncertainty of the average value for a given stadium, we have put "error bars" on these data. These tell us that if we were to accumulate data for another two seasons in a given stadium, there is approximately a two-thirds chance that the average distance for the new data would fall within the vertical limits of the error bars. Our

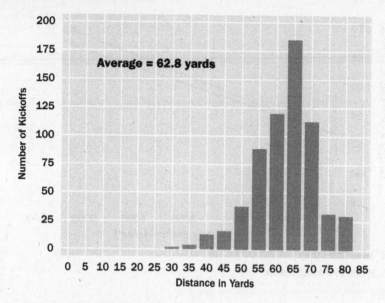

a) Sea-Level Kickers at Home

Average = 62.8 yards

Number of Kickoffs

Distance in Yards

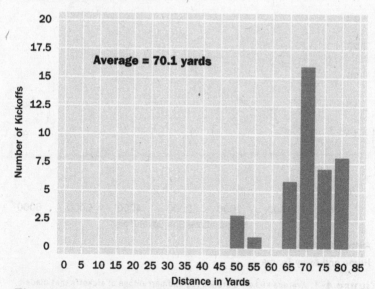

b) Sea-Level Kickers in Denver

Average = 70.1 yards

Number of Kickoffs

Distance in Yards

Figure 4-2. Frequency distribution of kickoff distances for a) sea-level stadiums and b) Denver's INVESCO Field at Mile High. The data are for kickoffs at/by Miami, Seattle, San Diego, Oakland, Washington, Baltimore, New York Giants, and New England during the 2001 and 2002 seasons. The average value for each data set is indicated.

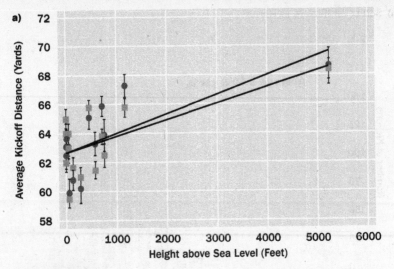

a)

• Away Team Only
▪ Home and Away Teams

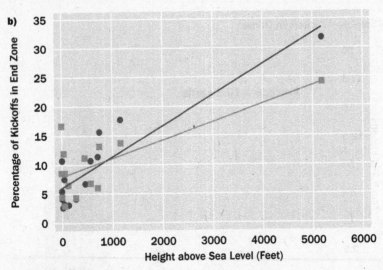

b)

• Away Team Only
▪ Home and Away Teams

Figure 4-3. Average kickoff distance (a) and percentage of kickoffs that made it into the end zone (b) as a function of stadium altitude. Data for all kickoffs (squares) and only those of visiting teams (circles) are shown. The two sets of data are not significantly different. Straight lines are determined mathematically to show the trends of the individual data sets.

Team/Location	Altitude (feet)	Fraction of nominal atmospheric density
Miami	10	0.9996
Seattle	10	0.9996
San Diego	13	0.9994
Oakland	42	0.9982
Washington	56	0.9976
NY Giants	60	0.9974
Baltimore	130	0.9944
New England	280	0.9880
St. Louis	455	0.9806
Buffalo	571	0.9758
Indianapolis	708	0.9700
KC	750	0.9683
Arizona	1,159	0.9514
Denver	5,183	0.8003

Table 4-1. List of teams that played in Denver during the 2000 and 2001 seasons, indicating the altitude (in feet) above sea level of their home stadiums, along with the atmospheric density at that altitude. For our purposes, altitudes of 500 feet or less are considered to be near sea level.

data suffer from being bunched at the low-altitude left end of the scale, but the trend is clear: kicking long in Denver gives you about a 5-yard advantage over kicking in Miami, the Meadowlands, or any stadium near sea level.

WINGING IT WITH WIND

Having gained some confidence in our air-drag analysis, let's apply it to answer two other frequently asked questions. First, what is the effect of wind on kicking (or passing) range? This can be a big issue at nondomed stadiums, especially in a place like Giants Stadium, where the wind seems to blow all the time, and at Lambeau Field, where recent renovations such as new luxury boxes and scoreboards have resulted in unpredictable wind patterns.

To simplify matters, we'll consider the case of a steady wind blowing directly down the field, either with or against the kicking team. The possibility of sideways forces due to crosswinds will be ignored, although including them does not greatly increase the difficulty of the problem.

The wind, of course, increases or decreases the horizontal drag on the ball, depending on which way it is blowing. The relevant speed for calculating this drag is no longer that of the ball relative to the ground, but of the ball relative to the air. The ball's vertical air and ground speeds are the same, but the ground and air speeds will now be different. If our kick has an initial horizontal velocity of 70 feet per second with a headwind of 20 feet per second, its horizontal air speed is now 90 feet per second, leading to greater drag. If the wind is following at 20 feet per second, though, the airspeed is reduced to 50 feet per second, with a corresponding reduction in drag.

The effect of wind speed on the range of the ball varies smoothly from the case of a strong headwind to a strong tailwind. If we consider our standard 45-degree tumbling kick at 100 feet per second, a headwind of 20 miles per hour at sea level reduces the range from 40 yards with no wind to 20 yards. In Denver, the equivalent range is reduced from 45 to 25 yards. For a tight spiral pass, the headwind ranges for sea level and Denver are 65 and 70 yards, respectively, compared with no-wind values of 78 and 82 yards. With a 20-mile-per-hour wind at our backs, the tumbling kick in the Meadowlands travels a distance of 62 yards, while the spiral pass in Denver will go 92 yards.

In **Figure 4-4** we have taken our standard 45-degree kick at

100 feet per second and considered the effect of a 20-mile-per-hour headwind or tailwind on the range of the kicker. Most affected, not surprisingly, are tumbling kicks at sea level, where the distance of a kick getting the push of a strong tailwind goes more than three times farther than a kick into the wind.

Obviously, wind can be a serious factor in a game. Not only can it reduce the range of kicks, punts, and passes, it can also affect accuracy, especially in stadiums whose configurations or locations encourage unpredictable swirling currents or unexpected gusts.

THE EFFECT OF TEMPERATURE AND HUMIDITY

Having considered the environmental factors of altitude and wind, let's take up two others: temperature and humidity. At a Miami

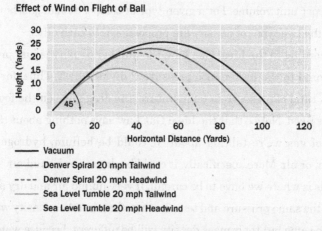

Effect of Wind on Flight of Ball

— Vacuum
— Denver Spiral 20 mph Tailwind
---- Denver Spiral 20 mph Headwind
— Sea Level Tumble 20 mph Tailwind
---- Sea Level Tumble 20 mph Headwind

Figure 4-4. The effect of wind on the flight of the football showing trajectories of tumbling motion at sea level and tight spiral motion one mile high for launch speeds of 100 feet per second and a launch angle of 45 degrees. Trajectories are shown for 20-mile-per-hour tailwinds and headwinds.

Dolphins game you might overhear the following excuse about a poor punt or kick: "It was so muggy that the ball just died." There appears to be a general perception that humid conditions significantly reduce the range of a kicked (or, in the case of baseball, batted) ball. How can we assess such claims? I think that with regard to humidity, such ideas follow from the basic idea of drag that we've just discussed. If the humidity is high, then there must be more water vapor in the air than normal, so the ball will have more molecules to run into and the drag force will be higher, right?

Wrong. In order to understand why this argument doesn't hold water, we turn to a new principle, the Ideal Gas Law of physics (and chemistry—but the less said about that, the better), which says that a gas's pressure, P, is proportional to the number of gas atoms per cubic foot—its "number density"—times its temperature, T. Now the quantity that's important for determining the drag force acting on the football (and hence its range) is its mass density, or pounds of mass per unit volume. For a given type or composition of gas, however, the two types of density are proportional to each other.

According to the Ideal Gas Law, the number density is just proportional to the barometric pressure P divided by T. Thus, for a given ratio of pressure to temperature, the drag force on the football is fixed. Notice that the Ideal Gas Law says nothing about the *type* of gas we're talking about. It could be helium, hydrogen, argon, or air. More specifically, it could be dry air or humid air.

This is where we have to be careful. If very humid air and dry air have the same pressure and temperature, their number density will be the same but their mass density will be different, because water molecules are lighter than the average weight of the different types of molecules that make up air. Thus, wet air at a given temperature and pressure actually has a lower mass density than dry air. It will thus cause less drag on the football. The fraction of water vapor is so small in even really moist air, though, that this effect is negligible. Temperature and pressure are the only variables that significantly affect the drag, and hence the range. As long as the thermometer and barometer readings are known, we can forget the hygrometer.

On the other hand, the Ideal Gas Law tells us that for a given barometric pressure, the temperature *will* have a significant effect on drag. Consider two football games: one played in Tampa Bay when the temperature is 100°F, and the Ice Bowl, where the temperature was −15°F. In order to use the Ideal Gas Law, we must convert these temperatures to the Kelvin scale, where zero temperature (sometimes referred to as absolute zero) means zero gas pressure. Zero degrees Kelvin is equivalent to −491°F or −273°C. At this temperature, the gas molecules have essentially no motion. On the Kelvin scale, 100°F is 311K, and −15°F is 247K. Thus, at fixed pressure, the density of the air decreases by a bit over 20 percent as the temperature increases from −15°F to 100°F. This is roughly equivalent to traveling from sea level to Denver, as far as air drag is concerned. Bottom line: Extreme temperatures can affect kicking distances on the level of a few yards. The ball will go farther in warm air at a given barometric pressure than it will in colder air at the same pressure. Humidity will have almost no effect on the ball's range.

"THERE ARE COACHES
WHO SPEND 18 HOURS A DAY
COACHING THE PERFECT GAME
AND THEY LOSE BECAUSE
THE BALL IS OVAL AND THEY CAN'T
CONTROL THE BOUNCE."

—BUD GRANT

CHAPTER 5

KICKING THE FOOTBALL

George Blanda was one of the greatest field-goal kickers of all time. He actually excelled as both kicker and quarterback over the course of his 26 seasons in the league, and he was productive to the end as the NFL's oldest player when he retired in 1976, just shy of his 49th birthday. Blanda played a key role in some of the classic pitched battles between the Oakland Raiders, his third and final team, and the Kansas City Chiefs in the late 1960s and early 1970s. Many of these games were decided by long field goals in the last minutes of play. The Raiders had Blanda to count on, but the Chiefs had a secret weapon of their own—tight end Morris Stroud. Morris stood 6 feet 10 inches and could jump high, even in pads. Instead of having him rush the kicker to block the ball, though, Kansas City coach Hank Stram would have Stroud lurk behind the goal to try to knock down any kick that was going to just barely make it through the uprights. Stroud never actually blocked a kick this way, but it was an interesting idea.

In chapter 4 we discussed the basics of projectile motion and the factors that affect a football's trajectory through the air. Absent, though, were the details of how the ball got moving in the first place. In this and the next chapter we'll look at the mechanics and tactics of the kicking and passing games.

To explore the physics of punting, passing, and kicking in more detail, we'll need to learn about angular momentum. Imagine a spinning merry-go-round. The ride itself isn't going anyplace, so it doesn't have any linear momentum. It is clear, though, that a lot of energy is stored in the merry-go-round because of its rotational motion. In the same way that objects with linear kinetic energy have linear momentum, the merry-go-round has rotational, or angular, momentum.

When motion occurs in a straight line, we consider force and momentum. But where there's rotation—of a merry-go-round or, for our purposes, of a ball or someone contacting the ball—then we must talk about torque and angular momentum. Just as force acts to produce linear momentum, torque applied to an object causes it to rotate and develop angular momentum. In football, the kicker's leg rotates about its hip joint and, after connecting with the ball, causes the ball to rotate about the kicker's hip joint as well as the ball's own center of mass. We will use these ideas to look at the mechanics of kicking.

THE MECHANICS OF KICKING

Let's begin with the kickoff. The kicker usually takes a good running start to build up speed. Whether he's kicking the ball soccer-style or straight on, he'll then plant his nonkicking foot to give himself a stable base and, swinging his hips, bring his kicking leg forward in a smooth arc, with a slight bend at the knee. This arc should begin as far back as possible. Just as the kicking foot makes contact with the ball, the kicking leg snaps straight, and the whiplike action increases its speed significantly. The end result is a collision between the kicker's foot and the ball. All of these actions, which must be practiced over and over to develop a smooth, fluid motion, have but one purpose: to maximize the launch speed of the ball, and hence its range.

To appreciate the mechanics of a kick from the physics perspective, let's first define some new terms by considering again the opening and closing of a door. If the door swings open a quarter of a full circle—90 degrees—in one second, we would say that it has an "angular velocity," ω, of 90 degrees per second. (Another unit of angular velocity you're more familiar with is "rpm," or revolutions per minute. Occasionally, we'll use radians per second, where a radian is 57.3 degrees. This is an odd unit, but it has the advantage that it makes calculations easier and equations simpler.) We'll use the Greek lowercase omega for angular velocity so that it doesn't get confused with regular linear velocity. The "moment of inertia," I, is the property of a body that tells us how hard it is to get it rotating for a given amount of torque. An important thing to remember about the moment of inertia is that it depends not only on the object's mass, but on how that mass is distributed. Imagine a door with a long, very heavy strip of lead running up and down the vertical edge near the handle. This added weight is going to make it more difficult to open. But what if we attach the same strip right next to the hinges? The door is now almost as easy to open as it was without any added mass. In this example, the moment of inertia of the door about its hinges is much bigger when the strip runs along its outer edge, even though the door's mass is the same in either case.

We now define the angular momentum, L, of an object around some axis of rotation. It looks just like a linear momentum, $p = mv$, except that we replace the mass (m) with the moment of inertia (I), and the velocity (v) with angular velocity (ω). Thus $L = I\omega$. In the same way, rotational kinetic energy $(K_{rot}) = (\frac{1}{64})I\omega^2$, since the kinetic energy of straight-line, or linear, motion $(K_{lin}) = (\frac{1}{64})mv^2$.

To calculate the speed of the kicked ball given the kicker's foot speed plus the speed at which he himself is moving, we need to look at angular momentum and the principle of Conservation of Energy. In chapter 1 we learned that we could calculate the speed of a player after a collision with another player if we knew their two masses and their speeds before the collision, and the speed of one player immediately after the collision. We could do this because total linear momentum doesn't change in any isolated collision between two players. For motion on a straight line, we can write $m_1v_1 + m_2v_2 = m_1v_1' + m_2v_2'$, where the primes refer to the speeds of the players after the collision. In order to calculate the final speed of player 2, though, we need to know the other three speeds in the collision.

In figuring out the speed of a kicked ball, we have an advantage:

in addition to conservation of (angular) momentum, the total kinetic energy of the leg plus the ball is also pretty much unchanged over the course of the collision. When energy is conserved like this, we say that the collision is "elastic." Collisions between players are usually quite inelastic. (Remember that the kinetic energy sneaks out in the form of heat, sound, and deformation.) We'll use the conservation of energy to give us another equation for the kicking collision, meaning that we will need to know only the initial speeds of the foot and the ball in order to get the final ball speed.

Just like linear momentum, the total angular momentum is always conserved in a collision. For the collision between the kicker's leg and the ball, we can thus write $L_{leg} + L_{ball} = L_{leg}' + L_{ball}'$, where, again, the primes refer to quantities after the collision. In this equation we are referring all of the angular momenta to an axis of rotation that passes through the kicker's hip joint. The ball's angular momentum after the collision is due both to its straight-line motion, and the fact that it may be rotating about its center-of-mass. The equation for the conservation of kinetic energy is almost as simple: $K_{leg,rot} + K_{leg,lin} + K_{ball} = K_{leg,rot}' + K_{leg,lin}' + K_{ball}'$.

The ball's kinetic energy is just the old $(\frac{1}{64})mv^2$, since it is simply executing straight-line motion. (We will neglect its rotational kinetic energy, which turns out to be small.) The kinetic energy of the kicker's leg is a bit more complicated because we have to take into account not only the rotational kinetic energy of the leg rotating about its hip joint, but also the fact that the hip joint is moving as well, giving the leg some straight-line-motion kinetic energy. These terms are labeled above as $K_{leg,rot}$ and $K_{leg,lin}$, respectively.

THE SPEED OF THE KICKED BALL

We're now ready to calculate the speed of the kicked ball using these equations. We want to consider the rotation and angular momenta of the ball and the kicker's leg about the point of the hip joint at the instant the foot and ball make contact. This physical system is shown schematically in **Figure 5-1**. In any kicking motion, whether it be a field goal or a punt, the kicker's hip is moving forward at some speed closely related to his center-of-mass speed. This motion of the hip is an important part of the fluid kicking motion. For a punter it is fairly slow, perhaps 3 feet per second. For a kickoff, it is closer to 15 feet per second. This motion causes a problem for our analysis because it means that the leg's pivot point is moving and must be accounted for by the $K_{leg,lin}$ terms in the conservation of energy equation, as mentioned above. (For purists, we must note that the lines of motion of the kicker's hip and foot aren't necessarily parallel. But factoring this in would make our calculations significantly more complicated, while not changing the answer very much.) In the meantime, the ball is sitting still on its tee, in the case of a kickoff, or being spotted in a fixed position by a holder in the event of a field goal.

We'll now employ a trick to make life easier. Notice that when we say the ball isn't moving or that the kicker is moving at 15 feet per second, we are taking the viewpoint of a fan sitting in the stands. Instead, let's hop on one of those camera gondolas that move on tracks up and down the sidelines during a televised game. At the instant of the kick, we'll arrange to be moving at a constant speed—say, 15 feet per second—so that to us the kicker isn't moving at all; his speed relative to the gondola is zero. Of course, this means that now the "stationary" ball is coming toward us at a speed equal to (but in the opposite direction!) what the fans in the stands observe the kicker to have. Looking at the kick from the perspective of the gondola simplifies life because our pivot axis for the rotations in the problem is now stationary in our new "frame of reference."

Let's now return to the specific case of a kickoff and consider our conservation equations from the point of view in which the kicker's hip is at rest. Remember that in the original frame of reference, where the fans are stationary, the ball is initially at rest and thus has no kinetic energy, but does have angular momentum about the moving axis of the kicker's hip. The leg has a complicated mixture of both linear and rotational kinetic energy. In the new frame, the

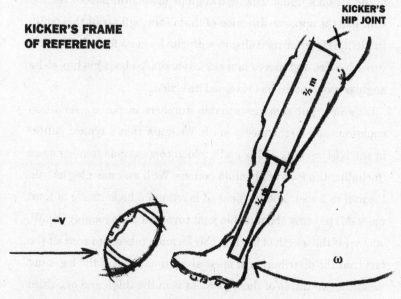

KICKER'S FRAME OF REFERENCE

KICKER'S HIP JOINT

Figure 5-1. Diagram of simple stick-and-ball model illustrating the mechanics of kicking. In the kicker's frame of reference, his leg rotates about his hip with angular velocity (ω) and the ball approaches from the left at velocity $-v$.

ball is moving backward (while still sitting on the tee) with velocity −v. The minus takes into account the "backward" part. For simplicity's sake, we'll make the reasonable assumption that at the point of contact, the kicker's leg is swinging in a plane that contains the ball, and that the launch velocity of the ball is tangent to the circular arc described by the kicker's foot (**Figure 5-1**).

What is the total angular momentum and kinetic energy of the system before the collision in our moving frame? The ball is moving counterclockwise about the kicker's hip (which is now stationary!). The kicker's leg now has only the kinetic energy of its rotation about the hip joint. In other words (and this is the main reason for going to all the trouble of changing our frame of reference), we can eliminate both K_{lin} terms from our conservation-of-energy equation in the kicker's frame. The leg's angular momentum about the hip joint has the opposite direction of that of the ball about this point. Immediately after its collision with the kicker's foot, the ball reverses course and moves in a clockwise sense about his hip, so its angular momentum has reversed direction.

Let's now put some reasonable numbers in our conservation equations and do the messy math. We know that a typical punter in the NFL weighs 215 pounds, which corresponds to a leg mass (including the foot) of about 35 pounds. We'll assume a leg length, l, equal to 3 feet. The moment of inertia of a kicker's leg of total mass (M) pivoting about its hip joint turns out to be roughly $(\frac{1}{4})Ml^2$, where l is the length of the leg. This formula takes into account the fact that the distribution of mass along the length of the leg is not uniform; two-thirds of the leg's mass is in the thigh and one-third is in the shin (see **Figure 5-1**). The official NFL Wilson football weighs 0.91 pounds. With all these variables plugged in we find the following result for the speed of the ball: $v' = 0.81v + 1.81l\omega$, where again, v is a positive number equal to the kicker's hip speed, and l is the length of the kicker's leg. To get the correct numbers from these formulae, we have to specify the speeds (v and v') in units of feet per second, leg length (l) in feet, and the angular velocity (ω) in radians per second. Finally, we must remember that this equation holds only in the TV gondola's frame of reference, in which both it and our kicker are stationary. To get back to the "fan in the stands" frame of reference, we have to add a term v to the equation's right-hand side. This makes sense if we think about it for a second: if the ball were stationary in the gondola frame, it would have a speed of +v in the stadium frame, so we get $v' = 1.81(v + l\omega)$. This formula neglects both the angular momentum and kinetic energy of the ball due to rotation about its own center-of-mass, both of which turns out not to affect our numerical results very much.

THE ACID TEST

Let's see how well this simple physical model predicts reality. Video analysis of typical leg speed in quasi-stationary kicks, whether they are done conventionally or soccer-style, shows that the end of the extended kicking leg is moving at a speed of approximately 60 feet per second relative to the hip joint just before contact with the ball. (This is essentially the kicker's ankle speed; foot speed is a less reliable measure of the leg's angular velocity because it is often rotating about the ankle before and during contact.) For a leg length of 3 feet, this corresponds to an angular velocity of 20 radians per

The Saints' Tom Dempsey prepares to boot the longest field goal ever—a game winning 63-yarder against the Detroit Lions in 1970. Note his foreshortened kicking foot.

second. Plugging this value of ω into our launch speed equation yields v' = 109 feet per second for a stationary kicker, or 74 miles per hour. This corresponds roughly to a punt, where the ball is not moving very rapidly in relation to the punter's hip joint when his foot makes contact with the ball. (The ball has about the same horizontal speed as the punter does, since he was carrying it before he released it.) If the kicker runs up to the ball from 10 yards away before booting it, as with a typical kickoff, he'll have a speed (v) of about 15 feet per second. This corresponds to a launch speed of 126 feet per second, or 93 miles per hour. The field-goal launch speed values will be somewhere in between those for punts and for kickoffs, because the approach distance is less than that for a kickoff.

How well do these numbers correspond to what we know happens in an actual football game? Given our assumption of conservation of energy, we might expect our model estimates to be a bit on the high side, since in reality some energy is always lost in these collisions. On the other hand, we have, for reasons of simplicity, neglected the effect of the lower leg "whipping" about the knee joint, which would increase the speed. Those two factors tend to cancel each other out, so we might expect our calculated speeds to be a reasonable estimation of the true launch speeds of the ball.

RECORD-SETTING
KICKS AND PUNTS

A good way to test our calculations is to analyze record-setting numbers—kicking at its extreme—and see how these compare with our estimates. Let's consider field goals first. The NFL record for a successful field goal is 63 yards, held by both the New Orleans Saints' Tom Dempsey (1970) and the Denver Broncos' Jason Elam (1998). The range of field goals is specified as the distance between the goalposts and the yard marker from which the ball is kicked. (The goalpost was moved from the front to the back of the end zone in 1978 by the NFL, but the field-goal distance of record is still specified in terms of the actual distance between the kicking tee and the uprights.) This complicates the geometry of the problem somewhat, because we're used to specifying the range of a kick as the distance between two points of equal height. The goal's crossbar is 10 feet above the ground. Since we are considering kicks with extreme range in this problem, we'll take the launch angle to be about 38 degrees. (It is difficult, if not impossible, to extract accurately from game films the actual launch angle of these kicks.) There is one crucial difference between the two NFL record field goals: Dempsey's was kicked in New Orleans, essentially at sea level, whereas Elam's was kicked in the old Mile High Stadium. This means that we'll have to account for differing air-drag factors. It also tells us immediately that Dempsey had to kick the ball harder than Elam.

What we want is a solution to the basic kinematics equations, including air drag, that give us a trajectory that passes just above the 10-foot-high crossbar a distance of 63 yards from where the ball is launched. Field-goal kicks almost invariably tumble end-over-end, so we'll use the appropriate drag coefficients for this situation. Given the known launch angle, then, we simply need to adjust our launch speed until we find the appropriate trajectory. Using our kinematics spreadsheet and the same drag coefficients we used in chapter 4, we determine that Dempsey needed to kick the ball with a speed of 173 feet per second, or 118 miles per hour. Elam would have had to launch his kick at 145 feet per second, or 99 miles per hour.

These are unrealistically high launch speeds, even for kickers with the talents of an Elam or a Dempsey. We are forced to conclude that either a) the drag coefficient we are using for tumbling motion is too high; b) the kicks that Dempsey and Elam made contained some component of spiraling motion, thereby reducing the drag coefficient; c) the atmospheric density was extremely low on the days Dempsey and Elam kicked their record-setting field goals; or d) a significant tailwind aided both kickers. In all likelihood, some combination of the first two effects is the culprit. No mention of tailwinds is made in game narratives for these two-kicks. Using our maximum calculated launch speed of 126 feet per second (remember, this is for a kicker running at 15 feet per second toward the ball), the average drag coefficient would need to be reduced by about 35 percent at sea level, and by about 20 percent at an altitude of one mile, to get the proper range for the same launch speed.

We know that a really booming kickoff will land beyond the end zone. Not much beyond, mind you, but it is reasonable to use 85 yards as an extreme in this case. Again assuming a 38-degree launch, an end-over-end tumble, and our drag coefficient reduced by 35 percent from the sea-level value that has been measured, we still get an unrealistically high launch speed of 171 feet per second, or 117 miles per hour. This again points to some spiral component of the kick, or the necessity of further reducing the drag coefficients from the reported values.

One final point about field goals: the NCAA record for Division I college ball is 67 yards. Why is the college record longer than the one for the pros? Shouldn't the skill level be higher for NFL kickers? Basically, the answer to this dilemma involves the statistics of large numbers. There are simply a lot more field-goal attempts in college than in the pros. In making a 67-yard field goal, one walks the fine line between athletic ability and luck. If your goal was to win the state lottery, you'd be better off buying 1,000 tickets instead of 100. Even though the average skill level in college is lower than it is in the pros, the much larger number of field-goal attempts means that someone, sometime, is going to get lucky and hit from way down the field.

It is difficult to make comparisons with long punts because the distances are recorded only for the total yardage gained, not where the ball first hit the ground. I've noticed that in college football, a very long punt through the air is about 75 to 80 yards. In this case, the launch speed is presumably about 100 feet per second, but if the ball spirals, that can reduce drag significantly. Assuming a spiral and a 38-degree launch at 100 feet per second, at sea level, we get a distance of 78 yards. This implies that the published spiral-drag data are pretty accurate.

Finally, we should note that our analysis involving the leg's moment of inertia addresses an important tactical consideration. All other things being equal, a tall kicker will kick the ball harder than a short one. This is because the angular momentum associated with the rotating leg will increase as the leg's length increases, assuming the mass is the same in both cases. From our equation for launch speed, and assuming that leg length increases in proportion with overall height, we can calculate that a 6-foot-4-inch field-goal kicker can launch the ball at a 13 percent higher speed than a kicker who weighs the same but stands 5-foot-7. For players of equal height, the more massive player will kick the ball farther if he can swing his leg as fast as the lighter player.

CALCULATING
THE FORCE OF A KICK

What is the force that must be applied to the football to get it moving at the high speeds we've calculated? We can estimate this by going back to our old, useful concept of *impulse* that we learned

Figure 5-2. High-speed stroboscopic photograph of a football kick. The duration of contact between the foot and the ball is roughly 8 one-thousandths. The deforming effect of the force of the kicker's foot on the ball, though brief, is obvious.

about in chapter 2. Remember that impulse is equal to the average force multiplied by the time over which this force is felt.

By taking high-speed movies of the exact instant of contact between the foot and the ball, we can experimentally determine the time interval over which the force is applied. A still photo (**Figure 5-2**) taken at very high speed (~1/30,000th of a second) makes it startlingly clear that the force must be very large to produce the deformation observed. In fact, the foot is in contact with the ball for about 8 one-thousandths of a second. Since the average force multiplied by the time interval over which the collision occurs (the impulse) equals the change in the ball's momentum, we can calculate the average force on the ball during the kick to be about 450 pounds. But this is just the average value. Starting at zero at the point of first contact, the instantaneous force the foot applies to the ball shoots up to several times its average value, perhaps reaching as much as a ton for an instant. No wonder the ball in the picture (**Figure 5-2**) looks like it's made of foam rubber!

ANGULAR MOMENTUM AND TUMBLING

We can also use what we've just learned about angular momentum to calculate how fast the kicked ball tumbles. The first and most important thing we need to realize about the tumbling ball's motion is that it is really a combination of two motions. The first is that of the center of mass: it makes a simple semi-parabolic path through the air, just like the ones we calculated in chapter 4. The second motion, added to the first, is that of the ball tumbling about its center of mass. This rotation means that the ball has angular momentum about its axis of rotation caused by the torque of the kicker's foot on the ball about its center of mass. Since the foot's force is generally applied below the ball's center of mass, this torque causes the ball to rotate in a counterclockwise direction (as we observe the kicker from his right).

How much tumbling angular momentum does the kicker's foot give to the ball? We can use the formula for impulse again, but in a slightly different form. Remember, every linear quantity we have studied, such as velocity or momentum, has a rotational equivalent—in this case, its angular velocity and angular momentum. Instead of saying that average force multiplied by the time interval of the collision equals the change in momentum, we now use the angular equivalent of this: average torque times the time interval equals the change in the *angular* momentum. This in turn equals the moment of inertia, I, times the change in ω. Now the torque applied to the ball about its center of mass equals the average force of the kicker's foot on the ball, F_{av}, times the lever arm that force has about the ball's center of mass. Let's call that distance d. The pros usually try to make d equal about 1 inch, just below the ball's center. We can use all these data to get ω, the ball's rotational speed.

Earlier in this chapter we mentioned that an object's moment of inertia depends not only on its mass but also on its size and the orientation of its rotational axis. The orientation issue now rears its ugly head. Here's a question for quarterbacks in the group: Is it harder to get a football spinning along its long axis, or about an axis perpendicular to the long axis going through the ball's middle? If you think about it for a minute, you realize that it's easier to spin the ball up along its long axis. The reason is that the ball's moment of inertia is smaller about this axis than it is about an axis that goes through the ball's waist. It's the same ball either way, of course—same mass, same shape—but now the axis of rotation has changed. When we do all the geometric calculations to figure these two different values of I, we get 0.081 pound-mass feet squared for the short axis, and 0.050 pound-mass feet squared for the long axis.

Putting all these data together and assuming a launch speed for the ball of, say, 100 feet per second, we calculate a value of $\omega = 93$ radians per second. Since there are 6.28 radians in a full circle, this corresponds to about 15 revolutions per second, or 900 revolutions per minute (rpm).

BAD TO THE BOUNCE

There are a number of reasons why kickers want the ball to tumble end-over-end. The first has to do with something we've already considered: conservation of angular momentum. When the ball moves off the tee, it has two kinds of inertia. There's the normal linear kind we learned about with the First Law. The mass of the football *wants* to go in a straight line. Of course it is dissuaded from following a straight-line path by the force of gravity. The second kind of inertia is the rotational kind. Conservation of angular momentum says that, unless acted upon by an external torque, an object will maintain its rate and orientation of spin. This tendency for a spinning ball to maintain its orientation is what stabilizes the tumbling ball in flight. This in turn improves a kicker's ability to accurately place the ball where he wants it to go. (More on this in chapter 6.)

What are the other options? If our kicker puts the arc of his foot right through the ball's center of mass, he won't exert any torque on it to make it tumble. You might think that this will make the ball go faster, but it won't. As long as the kicker delivers a given impulse to the ball, it doesn't matter where he kicks it—the center-of-mass flight speed will always be the same. (The reason a ball kicked near one end won't go very far is that it rotates out of the way before the kicker can really lay into it and deliver a big impulse.) Now the ball moves off the tee with no rotational motion. It thus has no angular momentum to stabilize it in flight. This means that it is very susceptible to air currents and temporary pressure instabilities that will cause it to fly through the air erratically, in much the same way a knuckleball does when thrown by a baseball pitcher. These small variations in trajectory can build up over time and significantly degrade the accuracy of the kick. On the other hand, the "knuckle" football can be used to good advantage when kicking the ball from deep in your own territory so that accuracy is not as important as distance and hang time. All Pro punter Craig Hentrich is famous for this kind of kick; it is very hard for the return man to figure out where the ball is going. This can throw off his timing, his catch, and his ability to get behind his blockers for the return.

The other problem with a nonrotating ball is that, leaving the tee with its long axis roughly perpendicular to its flight path, it experiences the maximum possible air drag. A tumbling ball will, over the average of a complete revolution, experience the average of the maximum and minimum drag forces. Drag can be reduced even more with a spiral punt, but this is next to impossible to pull off completely in placekicking. Pro punters can manage a rough spiral perhaps 80 percent of the time.

KICKING STRATEGY

We've developed a simple physical model to describe the mechanics of punting and kicking. Now let's revisit how kicking and punting strategies are affected by physics issues. First we'll take up what Peter Brancazio, another football physicist who has written several nice articles on the subject, refers to as the "kicker's dilemma." (Technically, we should call it the "punter's dilemma.") It is always good for a punter to maximize the punt's hang time in order to allow his team ample time to cover the punt. As we've seen from the discussion of chapter 4, though, maximum hang time is not compatible with maximum range.

The first question we want to ask is: When does a field-goal situation become a punting situation? Taking data from the 2001 and 2002 seasons in the NFL, plotted in **Figure 5-3**, it is clear that the odds of kicking a field goal when the line of scrimmage is farther than 33 yards from the goal line are dropping fast and have fallen below 60 percent. (This value assumes that the kick is made 7 yards behind the line of scrimmage.) Of course, the actual break point depends on the specific kicker a team has—and how crazy and/or desperate the coach is!—but for the sake of discussion, we'll call the 30-yard line the break point. It is interesting to note, in this regard, that the number of field goal attempts, as shown in **Figure 5-3**, drop off dramatically above 55 yards, much faster than the success rate is dropping. This is due primarily to the increasingly unpleasant consequences of a miss for the kicking team as the line of scrimmage gets closer and closer to their own goal line.

With the ball being snapped from the 30, the punter will want to pooch the ball into what's commonly known as the coffin corner: if the ball is downed there, the other team is dead. To bury itself in the corner, the ball has to travel 40 yards downfield and roughly 25 yards to either side, which corresponds to a kick range of about 47 yards. Using our calculated launch speed for a stationary kick of

109 feet per second, what launch angle does this correspond to? Well, it depends on how the ball is kicked—end-over-end or a spiral. We'll assume that we have a first-class punter, meaning that he can spiral the ball more than 80 percent of the time. Now, a punted spiral is not quite as aerodynamic as a pass spiral, so we'll use a drag coefficient that is one-third of the way from a clean spiral to an end-over-end kick. Putting these numbers into our kicking calculator and assuming sea-level air drag, we get a launch angle of either 17 degrees or 68 degrees. Both will give the punter the right range, but they give different hang times: 1.7 seconds and

Field-Goal Kicking Percentages

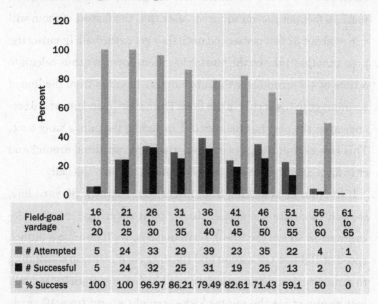

Field-goal yardage	16 to 20	21 to 25	26 to 30	31 to 35	36 to 40	41 to 45	46 to 50	51 to 55	56 to 60	61 to 65
# Attempted	5	24	33	29	39	23	35	22	4	1
# Successful	5	24	32	25	31	19	25	13	2	0
% Success	100	100	96.97	86.21	79.49	82.61	71.43	59.1	50	0

Figure 5-3. Field-goal kicking percentages as a function of kick range. The line of scrimmage yardage is 17 yards less than the kick distance. Data are for the 2001 and 2002 seasons for the teams listed in Table 4-1 on page 123, playing at home, and their visitors. The dark gray bars indicate the number of field goals attempted for the indicated range; the black bars show how many of these field goals were good; the light gray bars indicate percentage of attempts that were successful.

5.1 seconds, respectively (although it's a little odd to talk about the "hang time" of a punt with a launch angle of only 17 degrees!). The higher launch angle is the way to go. In addition to a longer hang time, 68 degrees has the advantage that it is harder to block; a 17-degree punt will hit the rushers right in the numbers.

How long will it take our guys to get downfield to cover the ball? They're spread out on the line of scrimmage within roughly 10 yards of the ball, either to the right or to the left. This means that on average they'll have to run about 40 yards to get to the coffin corner to cover the punt. Let's use our kinematic analyses of sprints from chapter 2 to figure out how long this will take. We remember that our fastest guys will fire off the line and accelerate for 2 seconds at almost 16 feet per second squared. After this, the fastest of them will run at about 32 feet per second until they get to the ball. Ignoring the pops that they take on the line to slow them down, we thus calculate a time of 4.8 seconds for them to run the 40 yards from the line of scrimmage to the corner of the field. The ball will land about 7.6 seconds after the play has commenced, including the punt's hang time. This means that the guys covering the punt can stand around and chat for 2.8 seconds before one of them has to down the ball.

If the punter is kicking from farther out than the 30-yard line, we can calculate the excess time a punt-covering team has depending on where the line of scrimmage is. We'll assume a launch speed of 109 feet per second and the drag coefficient corresponding to a modified punting spiral at sea level. For simplicity, we will just assume that the punter kicks straight ahead from 10 yards behind the line of scrimmage, wherever that may be. The distance our punt-coverage unit has to run is now just the straight-line distance to the goal. The longest our punt can go with this launch speed is 70 yards, corresponding (ideally) to a ball hiked from the punter's own 40. What we find is that if the line of scrimmage is farther than 55 yards from the goal line, the punt can't be covered

Punter Ray Guy, who inspired the term *hang time*, shows off his textbook form.

by even the fastest players. A longer punt will land before any of the offense can get there to down it. Thus, if a coach is facing a significant return threat—say, Dante Hall—he'd be wise to tell his punter to limit his range accordingly. As the line of scrimmage gets closer and closer to the goal line, the excess time that the coverage unit has increases. For a punt from midfield, there is an excess time of 1 second. This increases to 2.7 seconds at 40 yards and reaches a maximum excess coverage time of 3.4 seconds at our break point of the 30-yard line.

FOOTBALL IS A GAS

Now here's a really clever idea. If hang time is so all-fired important to the kicking game, why not fill the ball with helium? That would make it lighter so it would stay aloft longer, right? The Goodyear blimp certainly has an impressive hang time! Actually, this question was posed to some of the writers at *Sports Illustrated* back in 1993. To answer it, they got a punter from Auburn, Terry Daniel, to test the idea. Before I tell you what they found out, let's see what our kicking model and air-drag analysis tell us. This'll give us a rough idea of what to expect.

First, our basic notion was correct. Filling the ball with helium to the pressure required by NFL rules—12½ to 13½ pounds per square inch (psi)—*does* decrease the ball's weight, but only by about a third of an ounce. Notice that we're keeping the pressure the same. You might think that this would mean that the weight was the same too, but that's not how gases work. How they *do* work is described by the Ideal Gas Law, which we discussed in chapter 4. If we fill up the football to some pressure at a given temperature, the number density of the gas atoms or molecules is fixed. Before we go further, I should point out that helium gas is made up of single helium atoms, whereas air is mostly nitrogen molecules (two nitrogen atoms stuck together), along with some oxygen molecules (for breathing!) and a smattering of argon atoms. Since the Ideal Gas Law holds for any kind of gas to a good approximation, we'll get the same pressure with helium or nitrogen molecules as long as they have the same number of particles per volume at the same temperature.

To understand why this is true, we need to understand what causes pressure. The pressure in a football is equal to the force of the air molecules (or helium atoms) on any interior surface area, divided by that area. The pressure is the same at any point on the inside of the ball. If it wasn't, the gas would flow around until the pressure *was* equal. The force causing the pressure is due, in turn, to the molecules hitting the inner surface and bouncing off, just like Don McNeal exerts a force on John Riggins when he hits him and bounces off.

How fast are these molecules or atoms moving? It turns out that their speed varies like the square root of the ratio of the gas temperature to their mass. The temperature must be specified on the Kelvin scale so that the atoms or molecules come to rest only at absolute zero: $-477°F$ and $-273°C$. At room temperature, nitrogen molecules have a speed of about 1,400 feet (or the length of about four football fields) per second. An air molecule, on average, has about 7.3 times the mass of a helium atom, so for a given temperature it moves at one-third the speed. On the other hand, the air molecule's momentum is 2.7 times bigger than that of a helium atom. Thus, when it hits the inner surface of the football and reverses course, it exerts a bigger force.

So why, for a given density of atoms, doesn't the more massive molecule produce more pressure? Answer: because it doesn't hit the inside wall of the ball as often. The helium atom doesn't pack as big a punch, but it travels back and forth inside the ball more quickly, so it ends up producing exactly the same amount of force per area, averaged over time, as the heavier molecules do.

Filling both the helium and air balls up to the same pressure means that all of the outward forces trying to inflate the ball are the same in both cases, so that they "balloon out" in the same way, i.e., their shapes are identical. This means that the air-drag forces will be the same on each ball. And since the volume of each ball is the same, they must contain the same number of particles, be they atoms or molecules. But the helium atoms have only 14 percent the mass of the average air molecule, so the total mass of helium gas is 14 percent of the mass of the same number of air molecules. The football contains 200,000,000,000,000,000,000,000 (2 hundred billion trillion) atoms or molecules, but the total mass difference is only one-third of an ounce. This tells you something about how small the masses of the individual particles are.

The lower mass of the helium-filled ball has two consequences. The air drag on either ball depends only on its velocity (squared) through the air, so both balls, traveling at the same speed with the same orientation, will experience the same drag force. But the lower mass of the helium ball means that its deceleration due to drag will be more severe. We can understand this better by imagining a Ping-Pong ball and a steel ball bearing of the same diameter, both moving at the same initial speed through a tub of molasses. Because their initial speeds are the same, the molasses drag force on each of them is the same at the beginning. The Ping-Pong ball slows down much more rapidly, though, because of its lower mass. In addition, it turns out that our stick-and-ball kicking model developed earlier in this chapter tells us that the ball's launch speed depends very little on small changes in the ball's mass. What will the overall effect of this mass reduction be?

In the *Sports Illustrated* test, both the helium and air-filled balls were kicked 10 times. The average range of the helium ball was 57.7 yards, with an average hang time of 4.66 seconds. The air-filled ball had an average range 2.1 yards longer (59.9) with a *longer* hang time: 4.93 seconds. We can get a rough understanding of this result from the discussion above. What do our kinematic air-drag calculations predict? Let's assume that our punter can kick the ball with the "pretty good" spiral we discussed above, with a launch speed of 109 feet per second each time. We'll adjust the launch angle to give us the right range for the air-filled ball. Doing this, we calculate a hang time of 4.45 seconds for the air-filled ball, compared with 4.44 seconds for the helium ball. The helium ball's corresponding range is 59.1 yards. The launch angle required to get these numbers is 54 degrees. These results are in rough agreement with the *SI* test.

Part of the difficulty in reproducing the numbers exactly is our uncertainty about the actual air-drag profile of the kicks. The differences may also be due in part to the experiment not being repeated a sufficient number of times. Under scientific testing conditions, such as using a large number of kickers and correcting carefully for wind, we would probably get better agreement with the calculation. It is unlikely that the increase in hang time with the air-filled ball, a mere 10 milliseconds, could be detected within experimental error.

As another check on our calculations, we could adjust our kicking parameters to see what drag conditions are necessary with a launch speed of 109 feet per second in order to exactly reproduce both the range and the hang time the average air-ball punt that Terry Daniel made. This exercise yields a kicking angle of 60 degrees, with a 22 percent required reduction in the drag coefficient. The helium ball still travels 59.1 yards, with a hang time shortened by 10 milliseconds.

LOCATION, LOCATION, LOCATION

Up until now we've concentrated mostly on the range and hang time of the kicked or punted ball and how wind and altitude affect it. Now we want to consider accuracy. There are two kinds of accuracy: longitudinal and transverse, i.e., up-down and sideways on the field. The most important example of transverse accuracy is the requirement that a successful field-goal kick pass between the goalpost uprights as it cuts the vertical plane at the back of the end zone. Longitudinal accuracy comes into play when a coach wants his punter to have the ball roll or fly out-of-bounds near the goal line. Since it is extremely difficult to predict the direction of a bounce (see chapter 7), the safe play is to have the ball fly out-of-bounds as close to the goal line as possible. If the ball goes out-of-bounds between the 20-yard line and the goal line, the punting team gains a significant advantage—the closer to the goal line the better, of course. But given the vagaries of wind and other unpredictable factors, the risk that the ball will enter the end zone for a touchback also increases the closer the kick lands to the goal line. The best strategy seems to be to aim for the middle of the angular region defined by the sideline goal marker, the point of the punt, and the sideline 20-yard marker. This means that the ball will, on average, go out-of-bounds at the 10-yard line (**Figure 5-4**).

We can characterize the entire problem of accuracy in terms of the angular range a successful punt or kick can have upon leaving the kicker's foot. Consider the two angular ranges defined by the goal and 20-yard markers on the sideline for a punt, and the goal-

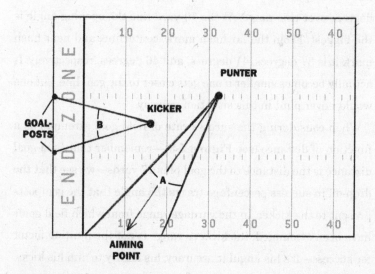

Figure 5-4. Definition of angles for punters to hit the coffin corner (A) and for field-goal kickers to be successful (B).

posts for a field-goal attempt (**Figure 5-4**). Starting 60 yards from the goal line (which would set a field-goal record for the NFL), the angle defined by the goalposts is 5 degrees, increasing to 9 degrees at the 30, and 17 degrees by the time we get to the 10. The opening angle of the goalposts is different depending on whether the ball is kicked from the center of the field or from one of the hash marks, but this difference doesn't become noticeable until we're right up on the goal line. There, it's 34 degrees from the center of the field and 32 degrees from one of the hash marks. Of course, as the angular range decreases, the probability of a successful kick or punt does also.

In the case of the angle defined by the goal and 20-yard lines along the sideline, there is little difference between punting from either set of hash marks as the punter moves from his own goal line to his opponent's 35. The angle increases from 4 degrees to 23 degrees over this range. At the 10-yard line the opening angle is the biggest. From the far hash mark, centerline, and near hash mark it is 37 degrees, 41 degrees, and 46 degrees, respectively. It actually becomes smaller if one gets closer to the goal line, but one would never punt in this situation anyway.

When considering the success rate of field-goal attempts as a function of distance (see **Figure 5-3**)—remember that field-goal distance is the distance to the goal plus 17 yards—we see that the drop-off in success percentage tracks the angle that the goalposts present to the kicker. In the yardage range from which field goals have been attempted, the kicker's range is not the limiting factor for success—it's his angular accuracy, his ability to aim his kicks. It is also true that in terms of opening angle, it makes almost no difference whether you're on a hash mark or the field's center line.

Indeed, the position of the hash marks in the NFL have crept progressively closer to the center line over the years. This improves kickers' scoring chances by increasing the opening angle of their target.

The field-goal data tell us something else as well. Occasionally, you'll hear about a clever coach who, when faced with a field-goal situation with the line of scrimmage in the red zone, intentionally takes a delay-of-game penalty in order to back his team up 5 yards, in the mistaken belief that this will increase the opening angle of the goalposts for his kicker. Actually, this was not a totally crazy idea in the old days of the NFL, when the hash-mark lines were wider than the goalposts. (This is still true in college and high school football.)

Think about one of the older wide hash-mark lines extended through both end zones. As we move up and down this line, the opening angle defined by the two goalposts is zero at two places: an infinite distance away from the goal and at the back of the end zone, where the hash-mark line intersects the line connecting the two goalposts. This means that somewhere along the hash-mark line there must be a point where the opening angle is maximum. And so there is, but it's in the end zone! In the case of an official NCAA field, this point is 12 feet beyond the goal line. For the NFL hash marks, the angle, starting with a value of 90 degrees at the back of the end zone, only gets smaller as you move away from the goal. Thus, taking the delay-of-game penalty to back yourself up reduces the chance of making a field goal. Believe it or not, there have actually been football coaches who never took a geometry or physics course in college.

There is one situation where it pays to take a delay-of-game penalty. If your punter is kicking a distance of, say, 45 yards consistently, and you're on the 41, it makes sense to back him up by 5 yards to minimize the risk of a touchback. As discussed earlier, the break point for punting instead of trying for a field goal is typically at about the 30-yard line. The punter's job is significantly easier than the field-goal kicker's at this line; the coffin corner has an angular range of about 25 degrees, compared with the 8-degree angle defined by the goalposts.

RABBIT'S FEET AND LUCKY TATTOOS

What can a kicker or punter do to improve his accuracy? Unlike linemen, who are a deliberative, logical, pensive group given to quiet self-introspection, these guys, as a class, are as superstitious as all get-out. Over the years, a large body of folklore has built up on ways to improve accuracy. Specific techniques of kicking, prescriptions for holders, and detailed requirements for shoes, not to mention the size and placement of rabbit's feet and lucky tattoos, are hotly debated. Several years ago, before the advent of the infamous "K" ball (which can be removed from its factory wrapper only in the presence of an official), *Sports Illustrated* ran an article on methods kickers use to doctor balls prior to games. The list included repeated inflation and deflation, immersion in ice water, microwaving, and virtually every other treatment (or mistreatment) imaginable except, perhaps, "sautéing . . . and plating them up with a nice port wine reduction." What works and what doesn't? What does physics have to say?

First, physics (and common sense) says: Hit the ball square. The ball will travel, at least initially, in the same direction as that of the force vector the foot applies to the ball. Let's say you want to make a 50-yard field goal from the left hash mark. This requires a directional accuracy of better than 7 degrees. Assume that you're kicking the ball straight ahead and that the force of your foot on the ball is perpendicular to the surface of the ball where contact is initially made. This means that the impulse delivered to the ball must be within three-eighths of an inch of the ball's centerline, or equator.

Such directional accuracy is more easily accomplished with a soccer-style kick—where the ball is kicked off the side of the foot—than with the straight-ahead style. Accuracy is accomplished in the soccer-style kick by angular placement of the foot perpendicular to the plane of the leg motion, whereas accuracy with a straight-ahead kick requires side-to-side spatial placement accuracy. When the kicker or punter is swinging his leg through a 90-degree arc with a terminal foot speed of 75 feet per second, it is much easier to orient the angle of his foot than to make sure the tip of his toe passes through a needle's eye with less than a half-inch radius. This simple physiological/geometric truth is responsible for the almost complete disappearance of straight-ahead kickers from pro and college football.

Before soccer-style kicking became popular back in the 1980s, placekickers would use shoes with squared-off toes to achieve the same effect. Perhaps the prime example of this was Tom Dempsey (see picture on page 138). Born with only half of a right foot,

Dempsey had a special shoe with a very broad, flat front surface. Jason Elam, on the other hand, who tied Dempsey's record nearly three decades later, kicked soccer-style.

SWEET SPOT? NOT

Punters and kickers spend a lot of time talking about kicking the ball at the point of its "sweet spot" to get longer, straighter kicks. This would be worth looking at if footballs had sweet spots. They don't. Sweet spots come into play in golf, tennis, and baseball with regard to club heads, rackets, and bats. Notice that the sweet spots of the *balls* are never discussed in these sports. (In football there could be a sweet spot on the kicker's shin, but he never kicks there.)

In physics the sweet spot of, say, a baseball bat is called the center of percussion and, unlike the center of mass, it is not a point that is defined uniquely by the geometry of the object in question. The bat's sweet spot is the exact point a ball must hit so the swinging batter feels no force at his hands when bat and ball collide. When the pitched ball strikes the bat, it delivers an impulse to the wood. If the batter's hands that grip the bat don't feel any force, Newton's Third Law tells us that they, in turn, could not have delivered any force, or counterimpulse, to the bat. This means that the total change of the bat's momentum is due to the force exerted on it by the ball. When we calculate the position along the bat at which the ball must hit it in order for this to happen, we find that the distance from the batter's grip to the point of impact is crucial in determining where the sweet spot is. The same is true for tennis rackets and golf clubs.

Because the football isn't being held by anything in the case of a punt, the concept of a sweet spot *per se* is meaningless. What about placekicking? The ball is being held, but not in a way in which the thing doing the holding—a tee or the ball holder's finger—can exert a significant force during the collision between foot and ball. There will never be a situation, except in the case of a very weird kick, where the tee imparts an impulse to the ball as it leaves the tee.

It appears that what kickers mean when they talk about the sweet spot is essentially the football's middle. For maximum range on a kick, it is certainly important to deliver the maximum impulse to the ball. This won't happen if the foot contacts the ball close to one of its ends, because the shoe (or bare foot) will tend to glance off the obliquely angled surface of the ball. The foot needs to connect solidly with the ball's bulk. The best place to kick a football is about an inch below its equator. This gives the ball the maximum impulse by lofting it upward at the proper launch angle, and also produces a stabilizing rotation. This means that the height of the foot above the turf must be adjusted depending on whether the ball has been placed on a tee for a kickoff or is directly in contact with the ground for a field goal or extra point. Kickers can adjust this foot height most reliably by keeping their kicking-foot swing the same, while adjusting the position of their "plant foot"—the one not doing the kicking. In the case of a field goal, with the ball on the ground, the plant foot is placed about 6 inches behind the ball. For a kickoff, the plant is usually made about a foot behind the tee.

When I was doing the physics of football segments for NFL Films, my producer, Brad Minerd, interviewed several NFL punters about the tricks they used. One player who shall remain nameless assured Brad that he was always very careful to punt the ball on the quarter-panel with the inflation valve. The sweet spot is there, he explained, because that panel has a second layer of cowhide to support the valve. These guys are pros, right? They've got to know what they're talking about. Very excited about learning this trick of the trade, Brad headed off to the Wilson football factory in Ada, Ohio, where all NFL game balls are made. When he asked the plant manager to explain how the second layer was sewn into the ball to support the inflation valve, all he got was a blank look. "We've *never* put a second layer in those balls," was the response.

THE PLACES FOR THE LACES

Technically, the football's laces, inflation valve, and seams all mess up the pure rotational symmetry of the ball about its long axis. In terms of placement accuracy, though, kickers need to worry only about the laces. Special team coaches always instruct their kickers and punters to kick with the laces out, facing away from the foot. The reason for this is clear: the laces make the ball's surface irregular, so the impulse delivered to the ball would have an unpredictable effect. Imagine trying to guess where a baseball covered with big welts would go when hit.

Some professional kickers have told me stories about kicked balls veering to the right or to the left depending on the initial position of the laces; they reason that the weight imbalance to one side or another will pull the ball to that side. This argument, at least in a first analysis, is wrong. Consider a dumbbell with one end weighing 10 pounds and the other 5 pounds. If the dumbbell is initially horizontal when dropped, it will remain horizontal as it falls. (Air drag will have minimal effect.) This lack of pull to one side is due to Newton's Second Law. Two objects with different weights fall with the same acceleration (and hence, speed) because the larger mass, which does feel a larger gravitational force, is also harder to accelerate. The counteracting effects cancel exactly.

If anything, one would expect the opposite effect for a kicked ball that tumbles about one of its short axes. If the laces are to the right, for example, there will be additional turbulence to the ball's right, which in turn will lead to a region of high pressure on that side. The higher pressure should push the ball to the left. Sudden gusts of wind can do odd things to the flight of the ball, and so can ball rotation. Rotational effects are often lumped together as being caused by "Magnus" forces, which are, for example, responsible for curveballs in baseball. They can also cause curving motion in kicked or punted footballs. We will take up the Magnus effect in the next chapter because it is for spiraling passes that such effects are most important.

"COACH,
HE WAS THE ONLY MAN OPEN!"

—CHICAGO BEARS'
QUARTERBACK
JOHNNY LUJACK,
AFTER THROWING
A THIRD INTERCEPTION
TO THE SAME DEFENSIVE
BACK IN ONE GAME

CHAPTER 6

PASSING THE FOOTBALL

Jim Plunkett was having a great year. The Heisman Trophy winner had been one of only four quarterbacks in the history of the game to be drafted first in the first round—yes, times have changed—but his final year with the Patriots and his two years with San Francisco had been marred by injuries and disappointing stats. Traded to the Oakland Raiders at the beginning of the 1979 season, he led his team to a wild-card berth the next year. Now, having beaten Houston, Cleveland, and San Diego in the play-offs, the Raiders found themselves in Super Bowl XV facing Philadelphia, whom they had lost to during the regular season in a squeaker, with the chilling knowledge that no wild-card team had ever won the Super Bowl before. Could they be the first?

Plunkett answered in the affirmative early in the second half when he completed his third and final touchdown pass of the day, this one to Cliff Branch. Dropping back from the Eagles' 29, Plunkett let fly a beautiful, tight spiral pass that arced smoothly through the air. At the end of its flight it appeared to nose to the right, veering into his receiver's cradling arms, and Branch stepped neatly over Philly rookie defensive back Roynell Young into the end zone. The Eagles never recovered, and Pete Rozelle was forced to hand over the Lombardi Trophy to one of his favorite people in the whole world, Raiders owner Al "Just Win, Baby" Davis, for the second time in five years.

Plunkett took home the Super Bowl MVP award for his performance that day, and Joe Montana collected the honor the very next year. The golden era of the quarterback had begun; in 1983 franchise players such as John Elway, Jim Kelly, and Dan Marino entered the league and with their high-powered, deep-ball offenses ran up incredible scores and stats, with Marino throwing for more than 5,000 yards in 1984 alone.

In this chapter we consider in more detail the part of football that has defined and revolutionized the modern game: passing. Of all the varied skills required of players, passing is the most difficult to master. This is one of the reasons quarterbacks get paid more than guards. Over the game's storied history, the football itself has evolved from what was essentially a soccer ball into a rugby ball and, finally, into its present form for two reasons: players wanted balls that were easier to carry and, as coaches began to realize the potential of the forward pass, they wanted in their quarterbacks' hands something that was easier to throw.

SPIN TO WIN

There are as many styles of throwing as there are quarterbacks—but what is the best way to throw a football? What effect does being right-handed or left-handed have on a pass? Does the quick release have a specific advantage dictated by physics? Where is the best place to hold the ball when passing? What about sidearm motion? These questions walk the line between physiology and physics; the answers depend crucially on both.

Ignoring short screen passes, shuttle passes, laterals, and the like, the basic goal of a passer is to throw the ball as accurately as possible with high speed, both to maximize range and to minimize the chances for an interception. There are two basic ways to achieve this goal. The quarterback could conceivably chuck the ball the way a soccer goalie throws, keeping his palm open and cradling the ball with its long axis in a semivertical orientation. He would swing his arm in a long arc, starting low from the back and increasing elevation as his arm moves forward, letting the ball roll off his fingers at the point of release. This throwing motion would cause the ball to spin on its vertical axis, with this axis being perpendicular to the direction of motion downfield. The second method is, of course, the one actually used: to throw the ball with an overhand or sidearm motion that imparts spin to the ball, again along its long axis, but with this axis pointing in the direction that the ball is being thrown.

The first method has both an advantage and a disadvantage. It allows the quarterback to apply a larger force to the ball over a longer distance than does the overhead motion; as more work is done on the ball, its kinetic energy and hence its launch velocity are greater at the point of release. The problem is that the ball is now in an orientation that causes maximum aerodynamic drag.

The overhead or sidearm throwing motion, on the other hand, causes the ball to spin so that it presents the smallest possible cross-sectional area to the onrushing air. Even with a shorter throwing arc, the best pro quarterbacks can throw the ball as much as 80 yards from a set position. Some quarterbacks, like John Elway, hold the ball toward its back. This allows them to apply more force along the ball's direction of launch, thus giving it greater velocity. Terry Bradshaw used to place his index finger at the back tip of the ball to provide even more force over his throwing arc. This can be particularly important for quick-release passers like Dan Marino, who work with an even shorter arm motion. Another ad-

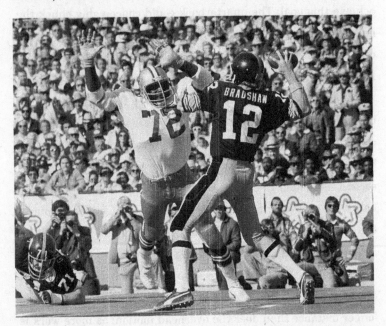

Pittsburgh quarterback Terry Bradshaw had a distinct style of gripping the ball—with his index finger on one point—that increased the speed of his passes.

vantage of the football-style throw is that it allows the quarterback to keep his eye on his receiver during the entire throwing motion.

Using our data for wind drag and assuming an optimal launch angle of 42 degrees for a tight spiral, a pass of 80 yards corresponds to a launch speed of about 101 feet per second. This isn't too tough for a quarterback with an arm like Darryl Lamonica. A football launched with a soccer-style throw at the same speed would go only 30 yards. Even if the soccer-style thrower could launch the ball at 300 feet per second (205 miles per hour!), the ball would go only 54 yards with its higher drag coefficient. Generally speaking, holding the ball at its midsection when throwing yields less launch speed but better accuracy and control, sometimes referred to as "touch." This can be crucial for slow, midrange passes where the ball has to find its way over a couple of defenders. Troy Aikman was never accused of having a cannon for an arm, but he was the master of touch.

Once the pass (or punt or kick) has been launched, how can we understand its wobbling, spiraling, or tumbling motion through the air? All of these involve the rotation of the ball about any or all of its axes, in addition to the basic motion of its center of mass in a semiparabolic arc. The rotational motion of the ball is a crucial factor in determining its accuracy and speed. A tight spiral pass can be a bullet to the numbers of a receiver in the end zone; a wobbling pass can be a duck begging to be picked off by a safety with a full head of steam going in the opposite direction.

The central concepts involved in this kind of motion are torque, angular momentum, and its conservation in the absence of torque. We've already touched on these in talking about blocking, tackling, and kicking. So far, though, we've considered only objects (footballs, linemen) rotating about a fixed axis. A door that is opening, for example, has an axis of rotation that is fixed in space and defined by the line joining the two hinges on which it turns. In order to deal with a football's motion through the air, we must now consider rotations in three-dimensional space—the axis of rotation can be wobbling all over the place. To do this, we must now start to think about torque and angular momentum as vectors instead of as simple algebraic quantities. This stuff isn't simple, but we can't fully understand the mechanics of passing or kicking without it.

TORQUE AGAIN

Up until now we have said that a torque is the simple numerical product of force and leverage. Now we want to consider both torque and the rotations it causes as arrows with a direction and a length, in the same way that velocity is a vector. Consider the force that we apply to a door in order to close it. Because we apply this force at a point removed from the hinges and perpendicular to the door's plane, a rotation of the door will result. What is the vector associated with such a torque? The short answer is that the torque vector points in the direction of your right thumb when you curl the fingers of that hand in the direction of the rotation that the torque tends to cause.

We will need to define things a bit more carefully, though. Let's consider the example of a force applied to the tip of a football (**Figure 6-1**). Remember that in order to cause a rotation, a force must be applied at a point away from the axis of rotation, and it

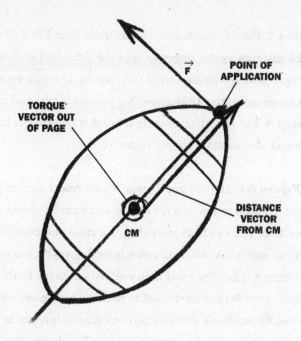

Figure 6-1. Torque applied to a football about its center of mass. The Right-Hand Rule indicates that the torque vector points out of the page. This is the same direction that your thumb points if you curl the fingers of your right hand in the direction of the football's rotation caused by the torque.

must point in a direction that does not pass through this axis. We want to consider rotation of the football, caused by torque, about an axis through its center of mass. We'll call the vector whose tail is at the center of mass and whose tip is at the point where the force is applied the *distance vector*. To find the torque associated with this combination of distance vector and force, we proceed according to a three-step process known as the Right-Hand Rule:

Step 1. Point the fingers of your right hand along the distance vector, from the axis of rotation to the point of application of the force.

Step 2. Rotate your fingers through less than 180 degrees from the distance vector direction into the force vector direction. If you're getting a twisted wrist from having to rotate your fingers through more than 180 degrees, flip your hand over and try again.

Step 3. The thumb of your right hand is pointing in the direction of the resultant torque vector.

In **Figure 6-1** (and in later figures) we represent vectors pointing out of the page at you as a dot centered in a circle; vectors pointing away from you will be represented as a cross and circle, like this: $\otimes$. (The idea is that when an arrow is coming at you, you see its tip, and when it is heading away from you, you see its tail feathers.) This torque vector is causing the ball to rotate in a counterclockwise direction. We can think of this sense of rotation as a vector as well: it's a vector corresponding to angular velocity. Which way does it point? Simple—just curl the fingers of your right hand in the direction of rotation so they point in the counterclockwise direction. The rotation vector is represented by your thumb.

Notice, now, a simple but crucial fact: both the rotation vector and the torque vector that caused it point in the same direction. Torque vectors point in the same direction as the rotations they produce! Now, remember that angular momentum, L, was introduced in chapter 5 as the product of an object's moment of inertia, I, and its angular velocity, ω. Not surprisingly, angular momentum is a vector as well, and it points in the same direction as the angular velocity vector with which it is associated. Thus, if I have an object that is initially at rest and I apply a torque to it, it will spin up, developing an angular velocity and a corresponding angular momentum. All three vectors—torque $(\vec{T})$, angular velocity $(\vec{\omega})$, and angular momentum $(\vec{L})$—point in the same direction.

So far so good. In fact, you may be wondering why we have to go through all these stupid rules to come up with something that is just common sense. The reason is that we need a careful vector definition of torque and rotation vectors in order to understand the physics of football rotation in three dimensions.

KEEP YOUR EYE ON THE BALL

Generalizing our discussion to include kicks and punts as well as passes, what does simple observation tell us about the motion of a football through the air? In the case of punting and kicking, which we took up in chapter 5, there are four basic kinds of motion. Kickoffs and field-goal attempts generally result in end-over-end tumbling, with the ball rotating about one of its short axes of symmetry. Such tumbling motion is usually pretty stable, in the sense that the ball's axis of rotation doesn't move about much—the ball doesn't wobble.

An ideal punt will involve a tight spiral, in which the long axis of the ball stays aligned (tangent) to the line of the ball's trajectory, i.e., the path described by its center of mass during its flight. This improves both hang time and range because air drag is minimized. Such punts are rare. A good punter, though, can produce a slightly wobbly spiral about 80 percent of the time. In these punts, the ball ascends rapidly with its long axis more or less parallel to the trajectory line. Toward the top of its trajectory, this kind of punt will do one of three things. Most commonly, the ball turns over (see **Figure 6-2a**) so that its long axis, after wobbling in a more or less vertical orientation as the ball reaches its maximum height, smoothly flips to a roughly horizontal and then nose-downward orientation in the vicinity of the trajectory's apex. Alternatively, the ball maintains its orientation over the entire course of its flight (**Figure 6-2b**). Finally, it can maintain its long axis roughly parallel to the trajectory line the entire time as a tight spiral (**Figure 6-2c**).

The majority of forward passes, on the other hand, can be classified as either tight or wobbly spirals, depending on how closely the ball's long axis tracks the trajectory line. (Any pass rotating on its short axis, end over end, is called a "duck" and would not qualify as professional at all.) Occasionally, the orientation of a forward pass will remain fixed over the course of the ball's flight, even though it may wobble just a bit.

THE CONSERVATION
OF ANGULAR MOMENTUM

How can we understand these different types of motion? Before we consider the effects of air drag that are required to explain the trajectories of **Figures 6-2a** and **6-2c**, let's begin by figuring out the simplest possible example: a pass (or kick) when air drag forces are not very important. This corresponds to **Figure 6-2b**, and is the

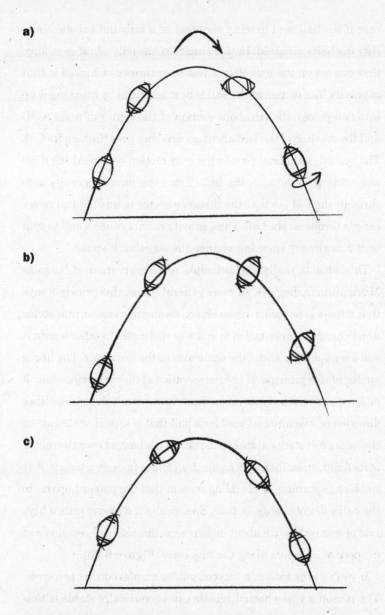

Figure 6-2. Various types of motion as the football flies through the air. a) A spiral punt that turns over. b) A pass or kick that maintains its long axis fixed in space. c) A tight spiral pass.

case if the ball isn't moving very fast or a tailwind substantially cuts the ball's airspeed. In this situation, the only significant force that can act on the ball after it has been thrown or kicked is that of gravity. The motion of the ball is best described by breaking it up into two pieces: the parabolic motion of the center of mass (CM) and the rotation of the ball about an axis that goes through its CM. The second, rotational part is the only motion we would see if we were riding along with the ball. Since the force of gravity acts through the CM (so that the distance vector is zero), it can never exert a torque on the ball. Thus, gravity cannot cause a ball to spin or, if it is already spinning, change the way that it spins.

This idea is really the principle of Conservation of Angular Momentum in disguise. In more general terms, this principle says that if there is no torque on an object, its angular momentum vector won't change its orientation in space or its length. In other words, it will keep spinning about the same axis at the same rate. The linear analog of this principle is the conservation of (linear) momentum: if no force acts on a body, its momentum vector won't change, in either direction or magnitude. Consider a ball that is kicked or thrown on the moon; its rotational state will remain unchanged over the course of its flight, since there are no air-drag forces to exert a torque. Any wobbling, spinning, or tumbling motion that the player imparts to the ball will not change in time. Specifically, if a passer puts a high rate of spin on the ball about its long axis, this axis will remain fixed in space at all points along the trajectory (**Figure 6-2b**).

If you've ever ridden a bicycle, you've employed this principle. The reason a two-wheeled vehicle can be vertically stable is that the wheels themselves act to store angular momentum. A bicycle traveling forward at a good clip has a considerable angular momentum (pointing to the rider's left) due to its wheels. Just like a spinning football, the wheels resist any change in their angular momentum vector, such as that due to a tipping motion. As the bike slows down and the angular momentum of its tires decreases, it becomes less and less stable.

This is also the physics principle behind gyroscopic stabilization once used for inertial guidance of rockets and submarines. In the days before accelerometers and global-positioning systems, you could determine your heading without reference to your external surroundings by using a very rapidly spinning gyroscope, mounted on low-friction gimbals. The low-friction mounts isolated the spinning wheel from external torques. Thus, its angular momentum, and hence its orientation, would remain constant over long periods of time, no matter what the rocket or submarine around it was doing. A motor attached to the gyroscope mount could be used to exert torque only in the direction of the wheel's angular momentum vector and thus keep it spinning without changing its direction.

TORQUE-FREE PRECESSION

Even if air drag is small, it is possible for the trajectory shown in **Figure 6-2b** to exhibit some wobble. Since we are still neglecting air drag, there is no torque on the ball, and its angular momentum vector, $\vec{L}$, is fixed in space. This doesn't mean that the ball's axes have to remain fixed in space. To convince yourself of this, take a football and throw it gently up in the air, to a height of maybe 3 feet, while putting a bit of spin on it about its long axis. The ball is moving so slowly that the air drag on it is negligible. Thus, $\vec{L}$ is fixed once the ball leaves your hand. Generally, though, you'll observe that the ball's long axis will wobble about a bit. This phenomenon is called "torque-free precession"; the ball's long axis is said to be *precessing* (not *processing*) about the fixed direction of $\vec{L}$. In fact, all three of the ball's axes of symmetry (the long one and any two mutually perpendicular axes in the "equatorial plane" of the ball) rotate about the fixed angular momentum axis. This is one of the things that makes "torque-free precession" so difficult to visualize; we are used to thinking about pure tumbling or spiraling motions, in which the full angular momentum of the ball is along one of its symmetry axes. In the case of torque-free precession, the ball is actually rotating about all three of its axes of symmetry. Quarterbacks can cause this complex rotation by launching it with a little flip or hook of the wrist just at the point of release.

Pure torque-free precession exhibits the interesting property that the ratio of the wobble frequency to the rotational frequency of the ball about its long axis has a definite value of 4/3, no matter how the ball is launched. This number depends solely on the shape and mass distribution of the ball. It can be calculated by doing some fancy math that we don't need to worry about here. You can verify the number 4/3 by taking a video of yourself lofting a few footballs into the air.

WHAT A DRAG (REVISITED)

Now we must grit our teeth and take into account the forces of air drag on the ball. These will be necessary to explain the trajectories shown in **Figures 6-2a** and **6-2c**. Such forces do not necessarily act along lines that intersect the ball's center of mass, so they can produce a torque on the ball, which in turn changes its state of rotation. We illustrate this idea in **Figure 6-3** on page 182. In **Figure 6-3a**, the ball's long axis is perpendicular to its velocity. The drag forces are all directly to the left and can be taken to act through the center of mass. In **Figure 6-3b**, the ball has a nonzero "angle of attack," and the air-drag forces vary over the surface of the ball. Because air molecules strike the ball in an asymmetric way, their total force on the ball acts along a line that generally will not go through the ball's center of mass. In the specific case illustrated in **Figure 6-3b**, the sum of all of the individual forces produce a net force that has both a drag component, pointing opposite the ball's velocity vector, and a lift component that tends to loft the ball above its trajectory line.

Let's consider the situation shown in **Figure 6-3b** in more detail. Things will now start to get a bit weird. Assume that a right-handed quarterback has thrown the ball with a good deal of spin along its long axis, and that it is ascending without wobbling along its trajectory, with a positive angle of attack—with the ball's long axis tilted upward relative to its line of trajectory. A right-handed quarterback will throw the ball in such a way that it is spinning clockwise about its long axis as he views it leaving him. (This spin-direction difference between right-handed and left-handed quarterbacks will turn out to make a

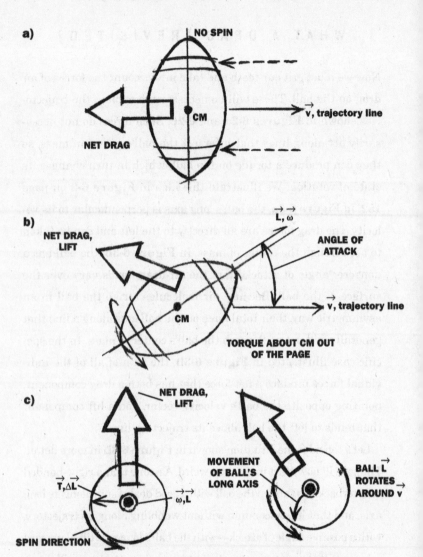

Figure 6-3. Forces and torques on a ball in flight. a) Ball's long axis perpendicular to its velocity vector. The net drag force acts through the center of mass (CM). b) Spinning ball has an angle of attack relative to its trajectory. The ball's angular momentum vector, angular velocity vector, and long axis are all parallel. The net drag force no longer acts through the CM. c) Looking backward along the ball's long axis with the same situation depicted in (b). Wind torque produces a small change in the ball's angular momentum, which causes the ball to rotate counterclockwise about its trajectory.

big difference in the flight of the ball.) Using the Right-Hand Rule (which has nothing to do with the fact that our passer is right-handed), we curl our fingers in the direction the ball is spinning, and our thumb points in the direction of both its angular velocity and its angular momentum. These vectors are indicated in the picture.

There are essentially two forces acting on the ball: its weight and air drag. The weight produces no torque. The air drag, which is the sum of all the little forces acting over the surface of the ball, acts like one big force applied along a line that passes above the center of mass. It thus applies a torque on the ball about the center of mass. We can figure out the direction of the torque using the Right-Hand Rule. Rotate your fingers from the distance vector (connecting the center of mass to the point where the line of the air-drag force intersects the ball's long axis) into the direction of the force vector itself. This will require that you move your fingers counterclockwise through about three-eighths of a full circle, or 135 degrees. When you do this, your thumb is pointing out of the plane of the figure.

One might reasonably expect the ball to tumble end-over-end in a counterclockwise direction as a result of this torque. *This commonsense idea is wrong!* The problem is that we've neglected the spin of the ball. Immediately after the ball leaves the quarterback's hand, its angular momentum vector, $\vec{L}$, is pointing in the direction of its long axis. The torque due to the air drag is pointing toward us, perpendicular to $\vec{L}$ and the plane of **Figure 6-3b**.

Figure 6-3c shows us how the situation appears looking straight backward along the ball's flight path. Remember that a force acting over a short time Δt causes a small change in an object's linear momentum. The same is true in the case of angular momentum and torque: $\vec{T}\Delta t = \Delta\vec{L}$. If we consider the effect of the air-drag torque acting for a short time, its effect is to add to the already present $\vec{L}$ vector by a perpendicular short piece $\vec{\Delta L}$ pointing to the left as we view the diagram. The new value of $\vec{L}$ is rotated counterclockwise in the direction of the torque caused by the air drag. The effect of this torque is not to cause the ball to tumble end-over-end but to rotate it to the left a bit (or, as the quarterback sees it, to veer right). As soon as the ball tilts, however, the direction of the force, and hence torque, changes. Now the drag force has a component in the direction the ball has moved because its angle of attack has rotated (**Figure 6-3c**). The new torque has a small component pointing downward, which causes $\vec{L}$ to dip a bit as well as continue to move to the left. The net result of this changing angle of attack, coupled with the air torque, is that the ball's long axis rotates (precesses) around its line of trajectory, as shown in **Figure 6-3c**.

You've seen this kind of motion before. If you set a top spinning rapidly on a floor, it will typically tilt from the vertical as its motion begins. The response of the top to gravity, however, is not to fall over but to precess with the tip of its spindle rotating in a horizontal circle. In the same way, the football doesn't start tumbling in response to the air drag; it rotates slowly about its line of trajectory.

END ZONE ANTICS

These ideas of torque and angular momentum can be illustrated by an interesting football phenomenon. Occasionally, when a player carries the ball into the end zone for a touchdown, he sets it down on the ground with its long axis horizontal and, in a fit of glee, spins the ball about its (vertical) short axis. (This generally occurs after he has called his mother on his cell phone.) The ball now does a very strange thing: it spins for a bit with its long axis horizontal and then "sits up," with its long axis now perpendicular to the ground. Why does this happen? When I posed this problem at a physics faculty meeting one day, the agenda was abandoned and the assembled group fell to bickering for half an hour about the answer. Had you been there, knowing what you know now, you could have easily explained it to them.

Let's assume that our guy spins the ball in a counterclockwise direction as he looks down on it. Thus the ball's $\vec{L}$ vector is pointing up (**Figure 6-4a** on page 186). When the player sets the ball down, it is essentially impossible for him to do it in a way that makes the ball's long axis perfectly horizontal. This means that as it spins, the contact point between the ball and the ground is not directly below the ball's center of mass. Thus, if the ball is tilting slightly up to the right, as in the figure, its point of contact with the ground is to the left of its center of mass. The frictional force acting at this contact point in opposition to the ball's rotational motion is into the plane of the diagram, because the ball, at least initially, is slipping and not rolling over the surface.

The effect of this frictional force, by the Right-Hand Rule, is to cause a small torque on the ball to the right and down. This torque in turn causes $\vec{L}$ to rotate clockwise just a bit relative to the ball's symmetry axes. We can split this rotation into two pieces. First, the spin of the ball about its short axis slows a bit, because the

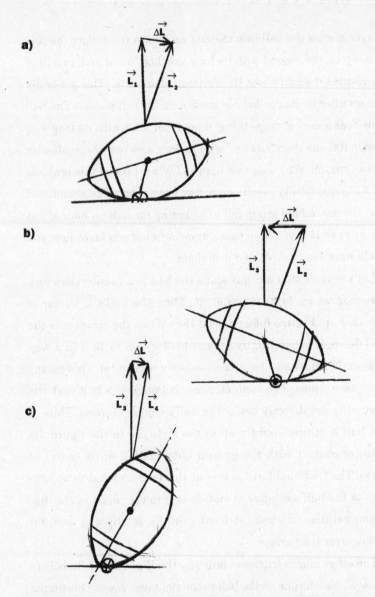

Figure 6-4. Torques cause a spinning ball to "sit up." The ball is spinning counterclockwise as it is observed from above. a) Frictional force between the ball and the ground (shown pointing into the paper) causes a change in the angular momentum from $\vec{L}_1$ to $\vec{L}_2$ (subscripts represent sequential times). b and c) This action continues as the ball rotates and pumps up the angular momentum along the ball's long axis.

vertical extent of $\vec{L}$ has been reduced. This is simply due to the frictional slowing of the ball. Second, the ball starts to spin about its long axis, because $\Delta\vec{L}$ points mostly in that direction. **Figure 6-4b** shows the ball a bit later, after it has rotated about the vertical axis by 180 degrees. The point of contact is now to the right of the center of mass, and the torque now acts to push $\vec{L}$ back toward the vertical, while at the same time continuing to shorten it in the vertical direction. Notice, though, that it is still acting to spin up the ball about its long axis.

This flip-flopping of the frictional torque direction gives us the key to understanding why the ball stands up. The $\vec{L}$ vector remains roughly vertical (while wobbling a bit) the whole time. But the ball itself is spinning faster and faster about its long axis; the torque and the ball rotation are perfectly synchronized so that spin along this axis is "pumped up" constantly. After a while (**Figure 6-4c**), most of the ball's angular momentum is along its long axis, while $\vec{L}$ is still vertical. This requires that the long axis of the ball be vertical as well.

At some point, depending on the amount of friction between the ball and the ground as well as how the ball is launched, the ball begins to spin so fast about its long axis that it starts to overrun the surface on which it is setting, and the directions of the frictional forces are reversed. The reversal of friction direction is analogous to a car wheel on the road. If the wheel is spinning slowly, as is the case when the brakes are applied violently, the frictional force is backward. If the driver is peeling out, on the other hand, the frictional force is backward. This reversal of frictional direction tends to make the ball wobble back to its starting position.

THE FLIGHT OF THE FOOTBALL II: DUCKS, TUMBLES, AND SPIRALS

We are now ready to understand all the different types of motion a ball can execute along its flight path. These are shown qualitatively in **Figure 6-2**, on page 177.

Case 1: Passing or kicking—negligible air drag (**Figure 6-2b**). This, as mentioned above, is the simplest possible case; we have already analyzed it, but summarize here the results. It occurs when the pass or kick is short, the launch speed is low, the wind and ball speed relative to the ground are roughly the same (thus minimizing the airspeed of the ball), or from some combination of these effects. The ball in this situation will do one of two things. If it is spinning about one axis, it will continue to do so over the entire arc of its flight, and that axis will remain fixed in space to conserve angular momentum. If, however, the ball is launched poorly, it will execute torque-free precession about its long axis. "Poorly" means that the quarterback or kicker puts spin on the ball around more than one of its axes. This typically happens with a pass if the player throwing the ball doesn't keep his wrist stiff as he releases it or fails to follow through. There is also more risk of this with a sidearm throwing motion, or if the player holds the ball toward the back during the throwing motion, like Bradshaw did. Wobbling motion can be caused by either a poor release or precession due to drag torque, which we'll discuss momentarily. It is relatively easy to kick the ball so that it rotates completely about one of its short axes, so that torque-free precession is usually not significant with tumbling kicks. Punters almost always produce wobbly spirals, in which the precession is more obvious.

Case 2: Passing—air drag important. We must now identify all of the air-drag forces and torques that act on the ball, as well as the possibility of torque-free precession that can occur whether or not air drag is important. A variety of analyses of this problem have been carried out in the last several decades by Professors Rae, Brancazio, and Soodak; the mathematically inclined reader is referred to the published reports on their work (see chapter 6 notes on page 272). What one can conclude from their extensive efforts, in addition to my own, is summarized here.

The air-drag force can be broken up into two pieces. The first acts along a line through the ball's center of mass in the direction opposite its velocity vector. The effect of this force was discussed in chapter 4, and essentially causes deviations of the center-of-mass motion from a parabolic trajectory. If the ball's angle of attack is nonzero and positive (negative), there will be an additional force component on the ball perpendicular to the ball's trajectory, causing it to be lifted above (or pushed below) the path it would have otherwise described. It is important to remember that the ball's long axis and its trajectory line (or, equivalently, its velocity vector) do not necessarily define a vertical plane. In the case where the ball's nose points to the left or the right of the trajectory line (sideslip), the air-drag force on the ball will have a horizontal component. These types of forces were not taken into account in the discussions of chapters 4 and 5. It is expected that, to the extent that the ball's angle of attack over the course of its flight *averages to zero*, these forces will cause little deviation from the paths and ranges we calculated. If, on the other hand, the ball consistently points its nose away from the trajectory line in one direction, the result will be a deviation from the pure drag situation in that case.

Unlike simple drag, angle-of-attack drag forces exert a torque on the ball as well, as shown in **Figure 6-3b**. This torque will cause the precession discussed earlier. Unfortunately, this precession in combination with torque-free precession tends to make the ball's motion very complicated.

Let's now break down what happens in the case of a very tight spiral (**Figure 6-2c**). This is the kind of pass that Jim Plunkett threw in our example at the beginning of this chapter. Such passes are often accompanied by an interesting phenomenon. When a right-handed quarterback throws the ball, it tends to veer a bit to his right; when a southpaw passes, it veers left. For the sake of discussion, let's assume we have a right-handed quarterback who launches the ball at an angle of 45 degrees. There are three main requirements for our pass to be a tight spiral: (a) that it have a high rate of spin about its long axis; (b) that the spin is only along this axis, so that initial wobble is negligible; and (c) that the ball be pointed along its trajectory, i.e., that its angle of attack be small. Well-thrown spiral passes typically have a spin angular velocity in the vicinity of 600 revolutions per minute.

Immediately after the ball leaves the quarterback's hand, its angular momentum and velocity vectors, $\vec{L}$ and $\vec{v}$, are parallel, and both are pointing opposite the air-drag force on the ball. Because of the ball's large angular momentum along its long axis, and with no spin about its other two axes, the ball's orientation in space remains fixed initially. As the trajectory line begins to curve down due to gravity, however, the ball develops a positive angle of attack. The first consequence of this is that a torque develops on the ball pointing to the right, as viewed by the quarterback (from whose point of view we will now describe all directions). As a result, the ball slowly swings with its nose to the right. A pass from a left-handed quarterback like Michael Vick or Kenny Stabler would experience a torque in the opposite direction and veer to the left. As the angle of attack increases due to the flattening of the trajectory near its apex, this precession becomes more pronounced. For typical launch speeds of 50 to 60 miles per hour and a spin rate of 600 rpm, the torque resulting from the developing angle of attack is sufficient to swing the tip of the ball to the right and down by 90 degrees over the course of the ball's flight. We can visualize this by plotting the "pitch" and "yaw" angles of the ball over the course of its flight (see **Figure 6-5** on page 192). These are the angles the ball's long axis makes above the horizontal plane (α; pitch) and the vertical plane (β; yaw) through its center of mass.

Over the course of the ball's flight, the drag-torque-induced precession has two important effects. First, it rotates the pitch angle of the ball down so that the long axis stays roughly tangent to the trajectory line. Second, it causes a yaw to the right, which results in air-drag forces pushing the ball to the right. This explains qualitatively why the ball veers to the right while keeping its nose close to the trajectory line.

Occasionally, a right-handed quarterback will release the ball with a *negative* angle of attack. In this case, the drag torque causes the ball to veer to the quarterback's left. As the trajectory line flattens out, however, the angle of attack gets smaller, rather than bigger as it does with an initially positive angle of attack, di-

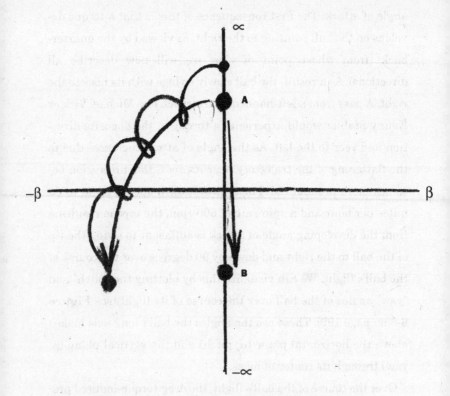

Figure 6-5. Pitch and yaw angles of the football's long axis as viewed backward along a horizontal line passing through the ball's center of mass. Positive/negative values of ∝ correspond to the ball's nose pointing up/down; positive/negative values of β correspond to the ball side slipping to the left/right as one looks backward. The position A represents the launch angle of the ball. For a tight spiral pass, with no side slipping, the ball's pitch/yaw position would move vertically downward from A to B. If the ball has no initial angle of attack, the precession indicated by the solid line occurs—a veering trajectory due to spin indicated by negative values of β. An initial vertical angle of attack results in the looping trajectory shown.

minishing the side-slipping tendency. As the ball descends, the angle of attack becomes positive, tending to bring the ball back right.

THE MAGNUS FORCE

In addition to these basic drag forces, if the ball is spinning we must consider the Magnus force, mentioned briefly in the previous chapter. The easiest way to visualize what causes this force is to consider a football that has been chucked so that it is moving horizontally but is spinning about its vertical long axis. Because the ball is spinning, one side has a higher airspeed than the other, so a greater drag force acts on it. This yields a net sideways force on the ball—the Magnus force. It is responsible for the curving action of baseballs and soccer balls. In the case of footballs, the effects are more subtle, as we shall see.

The Magnus force has a complicated effect on the trajectory. As the ball is ascending and developing a positive angle of attack, it tends to push the ball to the left, in the opposite direction of the air-drag yaw force. If the angle of attack becomes negative on the descending part of the trajectory, this effect is reversed, with both the yaw and Magnus forces acting to push the ball to the right. Indeed, careful observation of such passes reveals that during the ascent of the ball, there is very little deviation to the right. The deviation occurs during the descent. We can now understand this as being a combination of two effects. The most significant is the nature of the acceleration of the ball to the quarterback's right. Even in the absence of a counteracting Magnus force, the yaw force to the right occurs most strongly toward the end of the ball's trajectory because this is when the yaw angle is the greatest. Even if the yaw force were constant, its effect would be most evident at the end of the flight, simply because the distance moved due to a constant force accumulates as the square of time since the ball was thrown. Thus, most of the distance traveled under constant acceleration for a given time occurs toward the end of that time. The fact that the Magnus force acts in concert with the yaw force during the descent phase enhances the sideslipping effect at the end of the trajectory.

There is another Magnus effect that has apparently not been considered in the literature but that was pointed out to me by Professor Marianne Breinig of the University of Tennessee. It explains why the leading point of a tightly thrown spiral follows the arc of the pass closely (see **Figure 6-2c**), even under a broad range of launch conditions. The tendency for the long axis of a rapidly spinning ball to remain tangent to its trajectory cannot be explained fully by a simple angle-of-attack-induced precession. There must be some additional restorative force that always tends to reduce the ball's angle of attack. Such a force can be provided by a "spin-drag" Magnus force on the ball. Consider a ball thrown by a right-handed quarterback with a positive angle of attack. We observe the ball moving from left to right, standing to the right of the quarterback. For simplicity, we will consider only the drag on the two points of the ball nearest and farthest away from us. Were the ball not spinning, the velocity of these two points relative to the air streaming past them would be the same and opposite to the ball's velocity vector. The ball's positive angle of attack combined with its spin, however, gives the point on the ball closest to us a bigger air speed, and subsequently bigger drag. (This effect is enhanced due to the fact that drag force goes as the *square* of airspeed.) Two torques result from this difference in the drag force from the front to the back of the ball. One points backward and simply acts to slow the ball's spin rate about its axis. The second acts vertically downward and causes $\vec{L}$ to dip toward the trajectory line. This is the restoring torque we need to make the ball's nose follow its trajectory for a tight spiral. By carefully doing the vector analysis, we see that the Magnus drag on the ball will act to provide a restoring torque no matter which way the ball is spinning, and no matter how the angle of attack is oriented.

Let's assume that the quarterback manages to put all of the spin of the ball along its long axis as he releases it. This will preclude wobble due to torque-free precession. In this case, under what conditions will a wobbly pass occur? Imagine first that the ball is thrown with very little spin. The opening angle of attack still produces a horizontal torque, but now, because of the ball's lower angular momentum, the new horizontal angular momentum caused by this torque tends to manifest itself as a tumbling of the ball as well as a precession about the trajectory line. Generally speaking, both effects occur, but they are much less noticeable if the initial angular momentum of the system is large. (With a spinning top, if the spin rate is high, any off-axis force will tend to cause simple precession. If the spin rate is low, however, a transverse force will make the top wobble wildly and possibly fall over.) Now, because the angular momentum of the ball is no longer pointing uniquely along its long axis, the ball begins to wobble as in the torque-free case.

The wobbling will become more pronounced if the ball is thrown so that it has a large angle of attack at the launch point (see **Figure 6-5** on page 192). In this case the torque on the ball is immediately large. Even if the initial spin is big, some of the sideways torque will contribute to pitching the ball up, causing a rotation about its horizontal axis, and wobbling will occur, although it will be suppressed by the large angular momentum along the long axis.

Although the wobbling in these two situations is closely related to torque-free precession, it has been noted in recent research by Professor Rae at the University of Buffalo that with the effect of air-drag forces and torques included, the spin-to-tumble ratio of the football increases from the vacuum value of 1.67 to about 1.9.

FINALLY, CASE 3

Case 3: Punting and kicking—air drag important. In the case of end-over-end tumbling, air-drag forces and torques are chaotic and difficult to characterize. Their net effect, though, is to slow the ball's center-of-mass speed, as we discussed in chapter 5, and its tumbling spin rate. Since the torque exerted on the ball is essentially antiparallel to its angular momentum, precession and effects due to Magnus-type forces are small.

Things become more interesting for high punts that are wobbly (or tight) spirals (**Figures 6-2a** and **6-2c**). Such punts usually start out with a fairly small angle of attack and lose speed rapidly due to their steep ascent. This combination of factors tends to minimize the initial effects of aerodynamic torque, meaning that the ball's orientation in space does not change appreciably as it climbs to its maximum height. Due to its steep launch angle, however, a large angle of attack will develop by the time it reaches this point, often approaching or even exceeding 45 degrees. It is apparent from **Figure 6-3a** that aerodynamic torque is nil for angles of attack equal to 0 degrees or 90 degrees. It follows that the maximum aerodynamic torque will occur when the angle of attack is roughly midway between these values—more or less at 45 degrees.

Thus, as the ball begins to pick up speed on its descent, it has an angle of attack that tends to produce a maximum torque to the side, and hence a maximum precession rate. As the angle of the trajectory line drops sharply, precession through 180 degrees brings the ball into a nose-down configuration, which is responsible for the "turnover" frequently noted with high punts. The rapid changing of aerodynamic forces and torques can also induce rotation about the ball's short axes, leading to torque-free precessional wobble as well.

Another interesting effect that can occur on punts, pointed out to me by Ernie Adams, director of football research for the New England Patriots, clearly involves the Magnus force. If the punter if right-footed, the ball will spin in the same sense as a ball thrown by a right-handed passer. Thus for spiral punts that turn over (**Figure 6-2a**), the air-drag torque will cause the ball to veer to the punt returner's left. This is the same effect as for a tight spiral pass. On the other hand, if the ball fails to turn over (**Figure 6-2b**), it descends with its spin angular momentum vector pointing forward and essentially perpendicular to its trajectory line. The resulting Magnus force causes it to veer to the returner's *right*. Thus the punt returner must watch carefully the progress of the ball past the apex of its trajectory to ensure that he's underneath the ball for the catch. Failing to do this, as Adams notes, "can be the difference between a clean return and an adventure."

THE CREST OF THE STORY

We now shift gears a bit to discuss one other effect that can occur in some passes. Up until now, we have assumed for simplicity that the ball's trajectory began and ended at the same height. What happens in a pass play, though, if the field has a significant crest running down its center, with midfield being a yard higher than the sidelines? Some fields have this feature to improve drainage. Consider a quarterback throwing from the top of this crest to a receiver on a sideline. Or what if someone throws from the sidelines to a guy at the top of the crest? What is the effect on the timing of the pass relative to that for a flat playing surface? We know one thing for sure from the Conservation of Energy: the speed of the ball will increase if the quarterback is passing downhill and will decrease if he's passing uphill. This implies that the timing will change, but by how much?

We'll assume that we're doing the passing close to sea level and that the pass is a nice spiral, made at a zippy 75 feet per second. We'll also assume that the ball travels half the width of the field (26.7 yards) while either increasing or decreasing by 1 yard in height, to take into account the field's crest. (The height of the ball above the field when it is released and caught is, for simplicity, taken to be the same.) These conditions require a shallow pass of elevation 13 degrees and 17 degrees for the downhill and uphill passes, respectively. They yield final speeds of 68.3 and 65.8 feet per second, and flight times of 1.15 and 1.18 seconds, respectively. Since the equivalent flat-field pass has a final speed of 67 feet per second and gives a time of 1.16 seconds, we can safely conclude that field-height variation will not disrupt passing timing.

"GENTLEMEN, IT IS BETTER TO
HAVE DIED AS A SMALL BOY THAN
TO FUMBLE THIS FOOTBALL."

—JOHN HEISMAN

CHAPTER 7

GEAR

It's déjà vu all over again as the Bills are playing Dallas in the Super Bowl—and headed for a loss. This time it's Buffalo's first outing against the Cowboys, in Super Bowl XXVII before a crowd of over 100,000 people at the Rose Bowl. It's been a humiliating evening for the Bills. They're trailing 52–17 with 5 minutes left in the game. Backup quarterback Frank Reich, throwing in relief of the injured Jim Kelly, is desperately trying to keep alive a drive that started back at Buffalo's 47. With the clock ticking away, the Bills move the ball to the Dallas 31 and are getting up some momentum . . . until Reich gets popped by defensive end Jim Jeffcoat and coughs up the ball, which is scooped up by the other Dallas defensive end, Leon Lett.

Now here we need to be reminded how unfair life is for the poor lineman, be he on offense or defense. These fellows do most of the work and get none of the credit. Finally, however, Mr. Lett is convinced that his chance for glory has arrived. Escaping from the line of scrimmage with his trophy fumble, he chugs toward the goal line 65 yards away, tasting victory. Joy overtakes his soul. He holds the ball out for all the passing spectators to see—evidence of one final humiliation for the hapless Bills. In his euphoric state of mind, 5 yards from nirvana, he doesn't notice that joy isn't the only thing overtaking him—so is Don Beebe, the Bills' wide receiver. Beebe deftly taps the outstretched ball from the 290-pound Lett's hand. It bounces into and out of the end zone: touchback for Buffalo.

In this chapter we're going to concentrate on equipment, from the football itself—and how to hold on to it!—to the padding inside the helmet and a few helpful accessories.

THE CHANGING FACE (MASK) OF THE GAME

The game of American football has evolved in a number of significant ways. Offensive and defensive strategies have changed, perhaps most noticeably in the huge emphasis on the passing game. (Although it must be said that a lot of "innovations"—the 3-4 defense and the shotgun formation, to name just two—that supposedly occurred relatively recently in the NFL, were actually introduced many decades ago.) As we learned in chapter 2, the players themselves have been steadily increasing in size, speed, and athleticism. A third area of development has been in the equipment that football players use and wear. We've come a long way from the spherical footballs and canvas tunics of the 1870s.

It is important to understand, though, that each of these evolutionary changes affects the others. If guys the size and speed of Michael Strahan and Dante Hall were to face off as did the gridiron gladiators of old—wearing leather helmets with no face masks and only wads of cotton batting placed ineffectually over their shoulders—devastating injury and even death would be a weekly occurrence. Thankfully, since the advent of polycarbonate helmets in the mid-1940s, both pass defenders and pass receivers can keep their heads up during a hit without fear of decapitation. This is one of the factors that has led to a dramatic increase in the quality of the passing game.

So now let's consider in detail the gear that the team's equipment manager has on his shelves: helmets, pads, shoes, gloves, and, most important, footballs. How these things are designed and how they work have a crucial influence on how the game is played.

THE PROLATE SPHEROID— AKA THE BALL

The shape of the football—that of a prolate spheroid, to use a math phrase—has three obvious but important implications. First, it's easier to throw than a sphere. The consequences of this were discussed in the previous chapter. Second, the football is easier to carry than a soccer ball. It can be tucked firmly between the ball carrier's arm and rib cage in a manner that makes fumbling much less likely. Finally, the football's unique shape gives it an exceptionally erratic bounce that has unpredictable consequences over the course of a game.

Let's consider the bouncing issue first. When we say that a football bounces crazily, we really mean that its bounce is a lot less predictable than that of, say, a soccer ball. On a flat, smooth field, and with no spin, a soccer ball executes what is called *specular reflection* (**Figure 7-1a**). Specular reflection is what a laser beam does when it bounces off a mirror; the angle that the in-

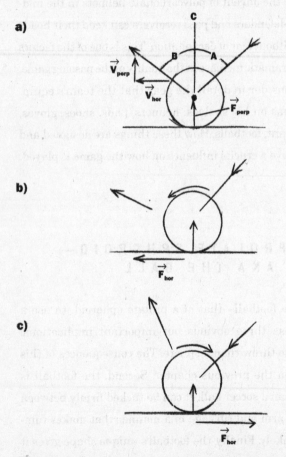

Figure 7-1. The bounce of a soccer ball. a) Without spin, the incoming and bounce angles (A and B, respectively) are equal (see text). b) Counterclockwise spin increases the bounce angle. c) Clockwise spin decreases it.

coming light ray makes with the mirror's surface, A, is equal to the angle going out, B. Consider the ideal case of a soccer ball that bounces without friction and without losing kinetic energy. In this case, the ground exerts a force on the ball that is perpendicular to the plane of the field. This force acts simply to reverse the component of the ball's velocity vector that is perpendicular to the ground, without changing its vertical speed. The ball's horizontal velocity doesn't change because there isn't any horizontal force during the collision. The bounce is completely symmetrical about line C.

In a real soccer-ball bounce, friction and energy loss come into play. Both of these factors make the ball's vertical rebound speed upward less than its initial downward speed. In the case of a spinless ball, since both velocity components are reduced in magnitude, the reflected angle may not be much different from the incident angle. Such effects are typically small, leading to a well-behaved if not necessarily completely symmetrical bounce.

Things begin to change with spin (**Figure 7-1b**). If the ball has a lot of forward (counterclockwise) spin, it will tend to leap forward. This is because the frictional force, which always opposes the relative motion between the ball's surface and the ground's surface at the contact point between the two, points to the left if the bottom of the ball is moving to the right. Similarly, a backward (clockwise) spin (**Figure 7-1c**) will cause the horizontal frictional force to increase to the right from its no-spin value. Thus, the ball's horizontal motion will be slowed down even more severely.

What is the combined effect of friction and spin? They essentially increase the range of possible bounce angles. The effect of friction is fairly limited, though. You rarely, if ever, see a soccer ball

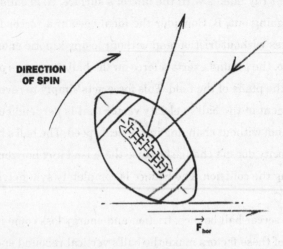

DIRECTION OF SPIN

$\vec{F}_{hor}$

Figure 7-2. Bounce-back of a rapidly tumbling kickoff. The horizontal forces acting at the football's point are large and to the right.

bounce backward, in the direction from which it came. For a ball coming in at 45 degrees, for example, you typically see bounces ranging from 25 degrees to 65 degrees in the forward direction, depending on the ball's spin and the condition of the field.

FOLLOW THE BOUNCING BALL

Now consider a football (**Figure 7-2**). If it lands with its long axis parallel to the ground, it will bounce in essentially the same way a soccer ball does. But a wider range of bounce angles are possible for a football. Much to the consternation of punt and kickoff returners, a football will land at their feet and proceed to bounce out of their grasp back toward the line of scrimmage.

The football's prolate shape also means that when it hits the ground with one of its points, a relatively large torque can be exerted about the ball's center of mass. In contrast, most of the force exerted by the ground on a soccer ball is the vertical force, which passes through the ball's center. This means it has no leverage on the ball, so it can't change its spin. The rotation of a football can change dramatically if one of its points is the first part to make contact with the ground. This change in the ball's tumbling motion can add significantly to the punt returner's difficulties, but it can also complicate the job of the offensive players trying to down the ball.

HOLDING ON TO THE BALL

Now consider a running back carrying the ball. On such plays, the main job of the defensive line and linebackers is to make the tackle and try to strip the ball from the ball carrier's grasp in the process. The running back's job, in addition to gaining as many yards as possible, is to hold on to the ball. Coaches will often require a fumble-prone back to carry the ball everywhere he goes—to class, to lunch, on dates—in hopes that a deep psychic bond will form between player and ball. Such mental drills may or may not prove effective. There are two physics-based techniques, though, that have stood the test of time.

The first is for the ball carrier to stay as low as possible. This was discussed in chapter 1. Presenting a small target to the defender and maintaining maximum stability on your feet by

Miami Dolphins' fullback Larry Csonka drags Minnesota tacklers toward the end zone during his MVP performance in Super Bowl VIII. Note the Zonk's secure four-point hold on the ball and the different styles of face masks being worn by players on both sides, including the horseshoe-shaped attachment designed to protect Csonka's nose, which had been broken countless times.

keeping your center of mass as low to the ground as possible make it less likely that you'll be tackled, and the less likely you are to be tackled, the less likely you are to lose the ball.

The second technique is the four-point method for carrying the ball. Let's assume that our guy carries the ball with his right arm. He holds the ball in such a way that it is supported at four places: his fingers curl around the front point, his forearm presses the front-right lower quarter against his right rib cage, his rib cage supports the rear-left upper quarter, and his right biceps presses against the rear tip. The four-point method makes it very difficult for a defensive player to hit the ball at any point, at any angle, and still dislodge it.

Let's go back to some of the ideas about torque and rotation we first touched on in chapter 1. Breaking the ball loose requires that the defender cause it to rotate about one of its points of support—by a big enough angle that it rotates right out of the ball carrier's grasp. But angular acceleration, and hence rotation, occurs only if there is an unbalanced torque acting on the ball. Consider a force that could be exerted by a defender ($\vec{F}_1$), shown in **Figure 7-3**. The force will cause a torque about the ball carrier's rib cage, causing it to rotate in a clockwise direction about that point of support. The ball carrier can apply three forces to counter this: at the front point of the ball with his fingers ($\vec{f}_a$); on the front quarter with his forearm ($\vec{f}_b$); and with his biceps, on the back point ($\vec{f}_c$). All three of these forces, due to their leverage, cause torques that make the ball want to rotate in the counterclockwise direction and hence oppose the clockwise torque applied by the defender.

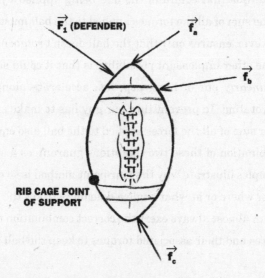

Figure 7-3. Four-point support technique to reduce the chance of a fumble.

Things can get more complicated if the defender's force has a significant vertical component as he tries to tear the ball free. Let's say a force is applied downward and to the upper back half of the ball. The resultant torque will try to rotate the ball downward about a horizontal axis passing through the ball's front point of support. This can be counteracted to some extent by the upward support of the ball carrier's forearm, but that force, being applied fairly close to the pivot point, doesn't produce a lot of torque; its leverage about the front of the ball is too little. The biceps and ribcage generally can't produce as much upward force, but the forces they *can* exert have much bigger leverage. The two countertorques will stymie all but the most severe blows delivered at this point.

Ultimately, our running back must ensure two different conditions. To keep the ball from rotating out of his grasp, he must apply a torque that counters the one being applied by the defender. The sum of all the torques applied to the ball must be zero. This, however, ensures only that the ball doesn't *rotate* out of his grasp. The other unpleasant possibility is that it could simply be pushed *linearly* out of his grasp, i.e., accelerate along a line without rotating. To prevent this, our guy has to make sure that the vector sum of all the forces applied to the ball also equal zero. The combination of these two conditions guarantees a safe tote. Our examples illustrate why the four-point method is so effective. No matter where or at what angle a defender strikes the ball, the runner can almost always exert the correct combination of two or three forces and their associated torques to keep the ball safe and secure.

HOW HELMETS WORK

After the football, the next most important piece of equipment on the field is the helmet. Unlike some of their predecessors, which were little more than leather skullcaps, today's technological marvels are remarkably successful in preventing serious head injuries. Basically, the modern helmet is a molded plastic shell that fits over the head, with a face mask and an interior lining of compressible material. The advent of face masks, and the replacement of leather with plastic, were both developments of the early 1940s.

Consider the hit that Buffalo Bills defensive back Mark Kelso put on Houston Oilers receiver Curtis Duncan in the incredible 1992 AFC Wild Card game. Duncan is in the end zone drawing a bead on what would be Warren Moon's third touchdown pass. Duncan himself, meanwhile, is being targeted by Kelso, who has built up a considerable head of steam. The ball arrives, and a split second later Kelso bashes his helmet into Duncan's, causing Duncan's head to fly back like a limp doll's. Fortunately, Duncan is able to pick himself up following the hit and celebrate the touchdown, his head still attached. (Unfortunately for the Oilers, they were about to blow the game in unforgettable fashion. Trailing by 32 points in the third quarter, the never-say-die Bills, again led by backup quarterback Frank Reich, put on a dazzling display of offensive fireworks to not only get back in the game but ultimately pull out the victory in overtime, 41–38, on a Steve Christie field goal. It would go into the books as the greatest comeback in NFL history—although the Oilers and their fans surely can be excused if they don't see it that way.)

Nothing could protect Duncan from the emotional whiplash he

would soon suffer, but how did his helmet manage to protect him from physical injury? We can answer this question by considering two physical quantities associated with a hit: pressure and impulse. We've already talked about impulse, and we'll return to it in detail in a moment, but let's first consider pressure.

Pressure is caused by a force applied over a given area. The actual value of the pressure is the force divided by that area: $P = F/A$. That's why we talk about pressure in units of pounds per square inch (psi). Remember that in chapter 5 we blew up a football to a regulation *pressure* of 13 psi. Things can get tricky here, though, because usually when we talk about pressure in this context, we really mean pounds per square inch as read by the gauge (psig), as opposed to an absolute pressure (psia). Absolute pressure is the pressure of the ambient atmosphere plus whatever the gauge reads. Atmospheric pressure, in turn, is what we feel as a result of the force of all the molecules in the air hitting our body. This pressure at sea level is roughly 15 psi. As the altitude increases, there are fewer molecules to hit a given area of our skin within a given time. The force per unit area is less, so the pressure decreases. If a football is blown up to 13 pounds, there are 28 (13 + 15) pounds of force pushing outward on every square inch of the inner surface of the ball.

When Kelso slams Duncan's head with his helmet, we can calculate the force of the hit by again using Newton's Second Law. In this case, Duncan's head and helmet, with a mass of roughly 20 pounds, accelerates to a speed of about 25 feet per second. The collision that causes this takes place in something like a tenth of a second. This corresponds to an average force during the hit of about of 160 pounds, but the instantaneous force can be much higher than the average value.

Now think about what would have happened if Kelso had kept his helmet on but Duncan had removed his. The momentum change (impulse) that Duncan's head suffered would be roughly the same with or without a helmet. It's fairly obvious, however, that without the helmet the result would have been catastrophically different. What saved Duncan's head?

Part of the answer lies in the fact that hard-shell helmets significantly reduce the pressure the victim's head feels. When it comes to injuries, the absolute force of the hit is not as important as the pressure that the force delivers. The crucial point here is that pressure is the force divided by the area over which it is applied. The force is distributed over the surface of the helmet facing the blow, instead of being concentrated in the area of initial contact between Kelso's helmet and an unprotected skull. The effective area of the collisional contact between the two heads is bigger with helmets because the helmet material is rigid and moves as a single unit to transfer the force, instead of bulging inward only in the region where the force is directly applied.

We can estimate how much the pressure is reduced by considering the ratio of the collisional contact area when Duncan wears his helmet to that when he doesn't. In the first case, the relevant area is about one-sixth the total outside surface area of the helmet. This is the part that actually ends up pushing on Duncan's head to accelerate it—the surface area under which the pads are actually squeezed so that they exert a force on his head. This area is perhaps one third of a square foot, or 50 square inches. If Duncan isn't wearing a helmet, there is direct contact between his skull and Kelso's helmet, with the area of the contact being closer to 4 or 5 square inches. Thus with helmets, the pressure of the collision is reduced by something like a factor of 10.

This pressure-reduction principle is the reason nails have pointy ends. If they had two flat ends, it would be very difficult to drive them into a board. A pointed end greatly reduces the area over which the force of the hammer blow is distributed, which provides a corresponding increase in the pressure that the hammer can apply to the board's surface.

To illustrate these ideas to my physics classes, I use a fun little torture device: the bed of nails. It consists of a ¾-inch-thick piece of plywood the size of a small bed, through which are pounded a couple thousand construction nails. The nails are 5 inches long and spaced about an inch apart on the board. The thing looks positively medieval. We put it up on the lecture table and I climb up and proceed to lie down on the thing, after which I continue lecturing. This demonstration makes the point very effectively. If I were to put all my weight on one nail by standing on it in my stocking feet, the pressure of that nail would easily puncture my skin. If I tried to do the same thing by balancing on two nails, one for each foot, the results would be unpleasant as well. As the number of nails is increased, though, my weight is distributed over more and more of these pressure points. By the time we get up to a thousand nails or so, the pressure from each has been reduced enough so that it can't do serious damage to my skin.

The nails have to exert a total upward force equal to my body's weight in order to keep the net force on me equal to zero, so I won't accelerate into the floor. With the bed of nails, though, the force, and hence the pressure, due to each nail is harmlessly small.

When I do this demonstration, I often am asked what the "trick" is. There is no trick—it's just the basic physics principle of distribution of force, and it is what helps football helmets do their job.

THE ROLE OF IMPULSE

The other physical principle that helmets utilize is that of impulse. Let's consider the helmet-to-helmet collision between Duncan and Kelso again. The force that each helmet applies to the other occurs over a short duration. We can graph this force, as shown in **Figure 7-4** on page 216. Here the force is plotted as a function of time. At the point in time when the helmets first touch, the force is zero. As the collision continues, the force ramps up to some maximum and then drops to zero again. The shaded area under this force-time curve is mathematically equal to the impulse, and the impulse equals the change in the helmet-plus-head momentum, or the product of their mass and velocity. The graph shown is the force of Kelso's helmet on Duncan's. According to Newton's Third Law, the graph of the force Duncan's helmet exerts on Kelso's would look exactly like this, except the force would point in the opposite direction.

To simplify things, we can pretend that the helmets just apply some constant force to each other over the duration of the collision. In this case, the impulse is equal to the area of a rectangle whose height is the constant force multiplied by the time over which the collision occurs. By choosing the constant force to be the

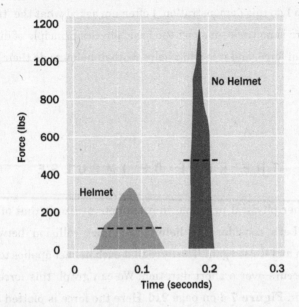

Figure 7-4. Force-versus-time curves for a helmeted and an unhelmeted head colliding with a helmet. The shaded areas under the curves are the impulses, or changes in momenta, and are the same in both cases. The average force during each collision is indicated by the horizontal dashed lines.

average force that the helmets apply to each other, we find that this rectangular area is the same as the area under the more complicated real force-time curve. We can thus talk about the impulse in terms of the collision time and an average force, without having to specify the complicated details of how the force actually varies in time.

Our understanding of impulse also allows us to see how the helmet's padding works. Consider the Duncan–Kelso collision once more. Remember that no matter what kind of headgear they're wearing, the momentum change—the impulse—their heads suffer will be approximately the same. In our first scenario, their helmets are on, but somebody snuck into the equipment room before the game and removed all the padding. Because the hard outer shells are still in place, we'll have the pressure reduction protection we talked about before. Without the padding, though, what the players feel after the collision is still bound to be unpleasant. Now let's put the padding back in. The violence of the hit doesn't change—its impulse remains the same. But this time the players survive, relatively unscathed. What's the practical difference?

Since the impulse isn't changed by the padding, the area under the force-versus-time curve has to be the same in both cases. But something has changed: the collision time. As the shell of the padded helmet is pushed toward its wearer's head, the padding compresses. This takes time, and the length of the collision increases (**Figure 7-4**). For the area to stay the same under the curve, the average force has to be lowered by a corresponding amount. It's this lowering of average force that saves both Duncan and Kelso.

The same idea is at work when we take two eggs and drop them from eye level to a concrete floor. Having been dropped from the same height, they have the same speed just above the floor. However, under one egg we have placed a 3-inch layer of foam rubber. Both eggs suffer the same impulse because both experience the same momentum change—they decelerate from their full speed down to zero. The one that lands on the foam rubber remains intact, though, because the average upward force that it feels from the floor is a lot less.

The impulse concept also applies to the webbed suspension helmets that were popular up through the 1970s. These helmets worked fairly well but didn't protect the sides of a player's head as well as modern helmets do.

A FOAMY HEAD BELONGS
ON YOUR BEER

The general physics principles that affect the game of football should be easy enough to understand—with a little professional help, of course. When it comes to applying them in a real situation, though, they're like power tools: care must be exercised. Back in the early 1990s, someone in one of the NFL's equipment rooms must have figured that if padding on the inside of the helmet was good, then even more padding, added to the outside, would be better. At the time, players were worried by a spate of concussions that seemed to be occurring. If you watch video of games from that era, you will occasionally see a quarterback or receiver wandering about the field wearing a big foam "crown" on top of his helmet. These guys looked like space aliens, but the concept did work up to a point. Players hit directly on the tops of their heads experienced less average force during the collision than guys without the added foam.

In some cosmic act of aesthetic karma, perhaps, this clever idea resulted in a worse problem, though. In addition to concussions, one of the most common serious injuries suffered by players is neck trauma due to the torque that occasionally gets applied unintentionally (or intentionally) to their face masks or other parts of the helmet. Helmets in the pros are often waxed to make them as slick as possible, making forces (hits) applied to the helmet by defenders more likely to glance off, minimizing torque and, hence, twists to the neck and head. With these goofy caps in place, though, there was a huge amount of helmet surface to be grabbed by any defender sailing past, and that increased surface was a lot "grabbier." The rate of neck injuries increased dramatically; as a result, you don't see this kind of padding anymore.

PADS, ETC.

As football players have become bigger and faster, equipment managers and sporting-goods manufacturers have come up with increasingly sophisticated and extensive protective gear. This has come primarily in the form of padding: shoulder pads, hip pads, tailbone pads, knee pads—you name it. Today's shoulder pads, marvels of mechanical design and materials science, have come a long way from the cotton padding players used to wear like epaulets. I sometimes tell my students that the history of physics goes hand in hand with humans' never-ending quest to kill each other more efficiently. That idea is illustrated quite well by the evolutionary development of football protective gear.

No matter how complicated it becomes, the basic idea behind padding is the same as we discussed with helmets: it lengthens the collision so that the average force felt by its wearer during a hit is reduced. This is particularly important for vulnerable hard-bone points like knees and tailbones.

Well, you may be thinking, if lengthening a collision is a good thing, why not put springs on players' uniforms? These springs would poke out at all angles, and the energy of a big hit could go into compressing the springs and not hurting the other guy. Brilliant, huh?

Well, springs *do* take a long time to compress and really would lower the maximum force felt by the players. True, all true. But as with foam helmet shells, one must be careful not to blindly apply the laws of physics. Let's see what practical problems might occur by modeling the players in a collision as two blobs of clay. When two blobs of clay, equal in size and mass and traveling with equal speeds in opposite directions, collide, they basically do what real football players do: stick to each other and stop. Momentum is conserved, but the organized kinetic energy of the collision is lost. A collision in which energy is lost is called inelastic. If we now put springs in the clay, so that the one on each blob is compressed when the collision occurs, the blobs will more or less rebound at the same speed they came in with. This is an elastic collision because little or no kinetic energy has been lost.

If you think about it, it becomes clear that if we put springs on players, something would be lost besides their dignity: control of the situation. If you hit a guy and he bounces off you, you've lost your ability to throw him to the ground or push him out-of-bounds, even though the forces you've experienced during the collision are relatively small. Regular pads soften the tackle, but it's still an inelastic collision; the players stick together.

One piece of protective equipment that doesn't work on the padding principle is a mouth guard. Being made of fairly hard plastic, mouth guards will not significantly extend the collision time between tooth and cleat. They act, instead, by the principle of distribution of force. A blow to a player's mouth will act on all the teeth at once, minimizing possible dental damage.

In the last several decades, gloves have become popular among players who need to catch or handle the ball. Even linemen and linebackers wear them, partially for protection and partially for ball control when necessary. Gloves are particularly good for receivers and defensive backs because they often employ naturally tacky materials to improve friction and, hence, the ability to hang on to the ball. Years ago, receivers and defensive backs would sometimes go to extreme lengths to take advantage of friction, literally coating their arms, torsos, and hands with "stickum." Fred Biletnikoff and Lester Hayes of the Oakland Raiders were notorious for this practice. The NFL outlawed it in 1981.

"FOOTBALL SHOULD
BE PLAYED ON GOD'S TURF.
PLAYERS SHOULD GET
THEIR UNIFORMS DIRTY,
EVEN A LITTLE BLOODY. . . .
THAT'S FOOTBALL."

—JOHN MADDEN

CHAPTER 8

TURF

The advent of artificial playing surfaces has definitely changed the game of football. For that reason, it has been the topic of extensive debate. There are whole Web pages devoted to the supposed effects of artificial turf on the performance of a team or even a specific running back. Players have their preferences, while manufacturers of artificial turf make their own claims. Sometimes turf even takes center stage on game day, as when conditions deteriorate after a downpour or snowfall on a natural-grass field, or when an artificial playing surface acts up, causing problems.

Take for instance the turf controversy that swirled around Philadelphia's Veterans Stadium at the start of the 2001 season. According to a report issued by the city's Office of the Controller in October of that year, "It was common knowledge long before [the summer of 2000] that the Veterans Stadium surface was considered one of the worst in professional sports, and that there was talk of legal action by the athletes who play at the Vet to force the city to replace the turf." But what to replace it with? (The offending surface was Astroturf, by the way.) Around that time I received several calls from the Philadelphia sports media seeking comment on the physics of artificial turf, including a product called NeXturf, which was laid down in time for the baseball Phillies to play their home opener.

Unfortunately, a pervasive drainage problem led to the cancellation of the football Eagles' own opener, a preseason game against the Baltimore Ravens, due to "spongy turf around the baseball cutouts." The problem was remediated with some modifications to the stadium's porous asphalt subsurface, but that proved to be a relatively short-lived situation. Construction on the Eagles' new stadium, Lincoln Financial Field—dubbed by the fans The Linc—had already begun, and the team played its inaugural game there on September 8, 2003, against the Tampa Bay Buccaneers on *Monday Night Football*.

And what of the infamous Vet? Like many of the older stadiums built before the heyday of luxury boxes and gourmet game food, in March 2004 it was imploded in spectacular fashion before hundreds of onlookers.

DOMES DOOMED GRASS

When domed stadiums made their debut in the mid-1960s, it was apparent that growing and maintaining a real grass field in them would be prohibitively expensive and difficult. Some sort of replacement would be needed. Grasslike carpets made of various plastics and fibers had been developed for indoor pools and playgrounds a few years earlier and were easily adaptable to the sports-field environment. Plus, artificial surfaces had an obvious advantage over grass: consistency. As coaches and players grew more and more systematic and scientific in the way they approached the game, they began to worry about the unknowns introduced by unpredictable flaws in a less-than-ideally maintained grass field. Groundskeepers enthusiastically supported the new artificial turf. So did team physicians, at least initially.

Early studies of Astroturf, conducted by none other than its manufacturer, 3M, purported to show that a smooth carpet reduced a variety of noncontact injuries that might befall football players. This seemed eminently reasonable; players were known to suffer ankle injuries from divots and other imperfections in grass fields. Installation of Astroturf and its variants, such as Tartan Turf and PolyTurf, surged in playing facilities across the country. By 1990 more than half of the stadiums in the NFL had artificial surfaces. Players found the surface more predictable, safer, and "faster." (Asking what is meant by "faster" turf produces almost as many answers as the number of players you question, but basically it means that plays seem to develop more quickly, and that a player can run faster and be more agile.)

After a brief honeymoon period, however, enthusiasm for the new fields began to cool. Players tackled on Astroturf might slide for 10 feet on their elbows and forearms before coming to a stop, an incredibly effective way to remove skin from one's body! In the middle of a long season, the concrete underlying the Astroturf seemed to get harder with every game, and running backs and wide receivers noticed the tendency of artificial turf to grab their cleats and not let go, just as they were planting their foot hard to make a sharp cut.

During the 1980s and mid-1990s, after a series of extensive studies, a consensus developed that playing football on Astroturf and its variants was, in fact, more likely to cause serious injury than playing on natural grass. More recently, a number of hybrid artificial playing surfaces have been introduced. The manufacturers claim that they truly are superior to grass in terms of consistency and safety. As of 2004, there are eight NFL fields with this kind of turf in use, two that still use Astroturf, and twenty-one that have real grass, such as Kentucky bluegrass or Bermuda sod. What does physics tell us about the differences between these surfaces, and what conclusions does it allow us to draw regarding their relative safety and speed?

GRASS BY ANY OTHER NAME

In the beginning, there was grass. Of course grass, in the context of football, can mean luxuriant Kentucky blue, parched bristle, sheet ice, swamp, or mud. That's the problem with grass. Its inconsistency produces unpredictable playing conditions. In addition to weather-related conditions, grass can have gouges, divots, and sinkholes. At the opposite end of the spectrum are products like Astroturf, which typically start out with an asphalt base, over which are layered plastic pads and a nylon mat bonded with nylon or plastic "blades" of artificial grass. Between Astroturf and real grass are the new "in-filled" composite products: FieldTurf, AstroPlay, and Sprinturf, among others, which use layers of increasingly fine crushed stone beneath a plastic porous backing bonded with long, plastic blades. The spaces between the blades are filled in with sand and/or rubber pellets, which serve as a cushion for running and keep the grass fibers from flattening against the mat. The rubber pellets are often made by grinding up old tires. At Memorial Stadium in Lincoln, the FieldTurf installation used 20,000 old tires from the dumps and yards of Nebraska. (This significantly reduced the mosquito population in the state, as well as improving our playing surface!)

While artificial turf requires significantly less maintenance than grass does, problems can arise. In the case of Astroturf-type surfaces, the plastic and nylon components are laid down like carpet, seams and all. Occasionally, older fields develop tears and bare spots. If the field on which the turf has been installed has dips or is very flat over a large area, water may fail to drain properly, causing puddles or ice patches. The newer infilled turfs must be carefully installed to allow for drainage and can also develop low spots.

THAT'S RESTITUTION

We can compare various playing surfaces in two ways: by measuring their springiness, or resilience, and by determining how much friction they generate. We'll take up friction in the next section and concentrate first on springiness. At one extreme, the surface could be made up of really stiff springs that would compress and extend just a little bit without absorbing energy and converting it to heat or sound. Such a surface would be very hard, like concrete or linoleum. At the other extreme would be thick mud. Anytime something with kinetic energy is thrown at this surface it will just sink in and stop, losing all of its energy. You've probably seen black-and-white photos or grainy film from the 1960s, before games were played on artificial turf or even decent grass, where the rain's coming down and the whole field is a quagmire. Anyone remember the 1965 Packers–Browns championship game in which the Pack's Paul Hornung was able to out-traction Jim Brown? Under such conditions, juking, making a cut, and even stopping is next to impossible, and the players' cleats get sucked into the ground on every step.

How can we quantify surface springiness? Since it seems to be related to energy loss in collisions with the surface, we can measure the energy loss directly and use this as a gauge of surface resilience. To this end, we define a surface's *coefficient of restitution*. If we take an object such as a basketball and drop it vertically from a height (H) onto a given surface, it will bounce back to a lesser height (h). The springier the surface or, equivalently, the less energy that gets absorbed by the surface in the collision, the closer h will be to H. The coefficient of restitution is defined as the ratio h/H.

We remember from chapter 2 that an object's total energy is the sum of its kinetic and potential energy, with its potential energy being equal to mgh, or the product of its mass, its acceleration due to gravity, and its height referenced to some zero point. When we drop our basketball, it has no velocity at first, so its energy is entirely in the form of potential energy. As it falls, its potential energy is converted into kinetic energy, but the sum of the kinetic and po-

Eagles' linebacker Chuck Bednarik looms over New York Giants halfback Frank Gifford after a devastating tackle on a muddy field during an old-school turf battle in Yankee Stadium on November 20, 1960. Natural grass seemed to suit "Concrete Charlie"; he missed very few games even though he played on both sides of the ball—offense and defense.

tential energies remains constant (neglecting the effects of air drag). When the ball hits the ground, it loses some of its kinetic energy. This new, lower kinetic energy is then converted back into potential energy as the ball rises, until it attains its new, maximum height. This height is less than the initial drop height; the ratio of the two is just the ratio of the total energies the ball had for its ascent and descent. Thus, the coefficient of restitution is a simple measure of the object's energy loss. For a perfectly elastic surface in which no energy is lost, it equals 1; for a mud pit, it is essentially zero.

Using coefficients of restitution to characterize surface properties is always a bit tricky because they depend not only on the surface itself, but also on what is being dropped and the height from which it is dropped. The coefficients of restitution for a given surface measured with basketballs are generally larger than those measured with foam-rubber balls of the same size. (Coefficients measured with eggs tend to be zero!) If a steel ball bearing was dropped onto a 1-inch-thick layer of mud covering concrete, the coefficient for a drop height of 2 inches would be a lot smaller than one measured for H = 15 yards. Thus, we must be extremely careful about inferring too much from a specific set of such data. Nonetheless, comparison of these coefficients can provide a rough measure of a surface's resilience and energy-absorbing capabilities.

There has been one study of this type published, in which the authors compared PolyTurf, Tartan Turf, Astroturf, and grass. (The details of the grass surface were not given.) A 16-pound shotput was dropped from a height of 6 feet onto the respective surfaces. This mass was used because it corresponds roughly to that of a helmeted 200-pound player's head. The coefficients of restitution were found to be 0.11 for the grass and PolyTurf samples, while Tartan Turf and Astroturf had coefficients of 0.17 and 0.20, respectively—showing that Astroturf is significantly more resilient than the other surfaces studied.

A more common method of measuring surface resilience is the "accelerometer" technique. This is employed by stadium groundskeepers who want to characterize the hardness of their fields, either from spot to spot or over the course of months or years to check for changes with time. One variant of this method involves dropping a massive "hammer," typically weighing 4.95 pounds, on the surface from a height of 18 inches. Embedded in the hammer is a force transducer—an electronic device that measures the force the turf is exerting on the hammer as it bounces from the surface. The parameter most often quoted from such tests is the so-called G-max value, corresponding to the maximum acceleration the hammer feels. (This is calculated by dividing the maximum instantaneous force by the mass of the hammer—Newton's Second Law once more.) This means that G-max data tell us about the maximum force (or "sharpness") of the collision, while the coefficient of restitution tells us about the total impulse the test mass feels, i.e., the average force it experiences times the duration of the collision. A high value of G-max will correspond roughly to a high coefficient of restitution.

For grass fields, G-max can vary widely; values between 30 and 170 are not unusual, with a value of 100 being fairly typical. (This highlights the variability of natural grass surfaces.) A value of 100 corresponds to an acceleration of 100 g's, or 3,200 feet per second squared. To put these values in context, remember that the maximum acceleration two players experience upon running into each other at top speed and coming to a dead stop is about 8 g's; the collision of the hammer with the ground occurs over a much shorter time, leading to the increased acceleration. For purposes of comparison, a frozen practice field may have a G-max greater than 300, while that of a tiled concrete floor approaches 1,300. Astroturf has typical G-max values in the vicinity of 110, while the new infilled synthetic turf products are usually closer to 90, although broad variations can be found in these numbers, depending on the age and condition of the field. Studies done at the University of Nebraska indicate that the G-max for in-filled turf is remarkably consistent over the course of several years, despite heavy wear.

FIGURING OUT FOOTING

Besides surface resilience, friction is the other major property that characterizes turf. We have talked about friction briefly with regard to open-field running and the use of gloves for catching passes, but its main importance for the game of football is in how it affects the playability of the field. There are four kinds of friction that are relevant to this discussion: sliding friction, static friction, torsional friction, and "torsional release" friction. Each type is characterized by a number, or coefficient, but all are determined by the details of how a player's shoes and/or his body interacts with the surface of the turf.

In order to understand how important these coefficients can be, think back—way back—to the 1934 championship game between the Giants and the Bears. Well-below-freezing temperatures and a hard rain the night before had turned New York's Polo Grounds into an ice-skating rink. Slipping and sliding around the field, the Bears managed to score the only touchdown of the first half, and led 13–3 early in the third quarter when New York brought out its secret weapon: basketball shoes borrowed from a local college! With a dramatically improved coefficient of static friction between their feet and the ground, the Giants scored a succession of unanswered touchdowns to beat Chicago, 30–13, in what is now referred to as the Sneakers Game.

With that slice of football history in mind, let's consider sliding and static friction first. When Dante Hall is making a cut during a kickoff return, he plants his foot so that it won't slip and pushes against the ground in the direction opposite that in which he wishes to go. If he pushes too hard, or if the ground is slippery, his foot slips out from under him and his run is over. The *coefficient of static friction* is defined as the ratio of the maximum horizontal force he can exert on the ground without slipping, $F_{breakaway}$, to the vertical (or "normal") force N that his foot is exerting on the ground. Mathematically, we say that $\mu_{static} = F_{breakaway}/N$, where we use the Greek letter *mu* for the frictional coefficient. The important idea here is that μ_{static} (like all the other frictional coefficients) is a constant; $F_{breakaway}$ increases in proportion to N. The coefficient of static friction is crucial in discussions of traction on turf and when one considers how much the turf grabs at the runner. Notice that we can rewrite the above equation in the form $F_{breakaway} = \mu_{static}N$: the breakaway force equals the coefficient of friction times the vertical force. Since the vertical force is usually (but not always) the weight of the player, this tells us that the bigger the player is, the bigger the frictional forces on him can be.

Once Hall's foot breaks away from its static grip on the turf and begins to slide, the *coefficient of sliding friction* comes into play. This coefficient is defined in a similar manner to that for static friction, except that the force in the numerator of the ratio is the frictional force on the *sliding* foot: $\mu_{sliding} = F_{sliding}/N$. Since this force is always less than the breakaway force for a given vertical force, the coefficient of sliding friction is always less than the coefficient of static friction. You can understand this by thinking about two strips of Velcro in snug contact. If you try to make the strips slide over each other sideways, the force required to initiate the sliding can be quite large, because each piece has its hooks firmly anchored in the other's "loops." Once the hooks have broken away, the sliding force becomes relatively small.

Friction between two objects, such as a shoe's sole and grass, works in essentially the same way. The microscopic bumps and "hooks" of any real object, no matter how flat it looks, dig into similar features on the other side of the interface, causing static friction. Once these features have broken apart and the two objects are moving relative to each other, the sliding friction is significantly less. Interestingly, the coefficient of sliding friction depends very little on the sliding speed.

Both the coefficients of sliding and of static friction refer implicitly to motion in a straight line. Imagine now that a player has placed his cleated foot firmly on the ground and is trying to rotate it. Because his cleats have dug into the turf, the initial torque he applies to his foot through his knee and ankle may be insufficient to cause the foot to break away. Similarly, once his foot has broken away, it will require a given amount of torque for the foot to continue to rotate. The maximum torque a player can apply to his foot without it beginning to rotate defines the torsional breakaway coefficient $\lambda_{static} = T_{breakaway}/N$. Similarly, the sliding torsional friction coefficient $\lambda_{sliding} = T_{sliding}/N$. The lambda coefficients are similar to the μ coefficients in that they depend crucially on the details of the shoe-ground interface and, to a good approximation, do not depend on N.

Clearly, shoes with long, sharp cleats will yield bigger coefficients than flat-soled shoes. Cleats—the ultimate in roughness— are just an extension of the general idea that rough surfaces sliding over each other have more friction. Pro football teams and most college teams have on hand a variety of shoe types, depending on the field to be played on and the anticipated weather conditions. Pepper Burrus, the Packers' head trainer, has said, "Our equipment people are like an Indy Pit crew. They can change cleats in the blink of an eye if conditions change."

Comparing the available data on the various frictional coefficients between shoes and turf surfaces is difficult because they have been taken under a variety of conditions, with significantly different types of shoes. Generally, researchers have found that μ_{static} and $\mu_{sliding}$ are roughly equal to 1, although great variations in these numbers are observed. This means that for a 250-pound linebacker, the maximum static frictional force the ground could apply to him would be 250 pounds. If both of his feet are firmly planted on the ground, this would correspond to 125 pounds of frictional force on each foot. If he was balanced on one foot, all 250 pounds would act through that foot. Similarly, most researchers find torsional coefficients to be about 0.8 feet. (Torsional friction coefficients are specified in feet because the measured frictional torque in foot-pounds is divided by the force in pounds that presses the shoe to the turf.) Thus a 200-pound running back, who plants one foot firmly on the turf and through which he supports his full weight while making a cut, requires about 160 foot-pounds of torque (applied through his ankle and knee!) to rotate his foot.

Despite the distressing variety of experimental conditions and methods to be found in the literature, we can still draw some general conclusions from the broad range of available data. The first and most obvious is that cleat length and configuration significantly affect the coefficients of friction. This is particularly true of the kinetic coefficients of friction, which tends to exhibit larger variations as turf, shoe types, and surface condition (wet versus dry) are varied. The effect of wet surfaces is to lower the friction coefficients by 30 percent to 40 percent from the dry values. One exception to this occurs when shoes are "spatted"—covered with athletic tape to simulate the practice of extended ankle wraps.

Grass and infilled turfs have roughly the same coefficients of sliding rotational and linear friction, with Astroturf being about 30 percent higher on average for both types of coefficients. It should be noted, however, that some researchers find grass to have higher rotational and linear coefficients for some shoes when compared with Astroturf. Interestingly, researchers who have measured static coefficients for torque and linear force find them to be only slightly larger than their moving, or dynamic, counterparts. There is some evidence that the torsional coefficients increase as the temperature of the turf increases, but this is not a very big effect. When artificial surfaces are warm and dry, they are particularly grabby.

THE FASTEST SHOW ON TURF

How do the differences among field surfaces actually affect how football is played? We'll consider three factors: injury rate, simple straight-ahead speed, and maneuvering ability, or the ease with which one can change direction or speed.

Studies carried out at all levels (see notes for chapter 8 on page 273) have shown that both contact and noncontact injury rates are higher for poorly maintained grass surfaces and Astroturf than they are for well-maintained grass or infilled turfs, particularly for knee and ankle injuries and ailments like sliding abrasions (turf burn), and concussive injuries like spurs, turf toe, and tendinitis. There is also increasing evidence that, on average, infilled surfaces are somewhat safer than grass, but no long-term studies have been done on this yet. In the case of contact injuries, a recent study has shown that there is long-term evidence for increased rates of head trauma in collisions between helmeted heads and Astroturf.

How can we understand these results? Consider ankle and knee injuries, primarily anterior cruciate ligament (ACL) injuries, first. These can be caused when the player's body is moving one way and his foot has failed to break away to follow the rest of his body. Forces and torques then build up in the flexible parts of his leg—the ankle and knee, and specifically in the muscles and tendons that support these moving joints. This immediately points to the coefficients of static friction and torsional release as the culprits in causing such injuries. High friction is also responsible for turf toe, in which the player's foot is repeatedly jammed forward into its shoe after the shoe is grabbed by the turf. Torg (see chapter 8 notes) has gone so far as to quantify the relationship between the coefficient of friction and safety; he claims that any value of $\mu_{static} > 0.4$ is dangerous. Turf burn is less a function of a high coefficient of friction than it is of abrasiveness. Grass and FieldTurf, which uses relatively soft plastic for its blades, are less abrasive. Grass and infilled turfs have their own issues. Unsure footing and ankle rolling can result if a surface is too soft. Grass, of course, can cause injury with its imperfections; a divot can blow out a back's ankle just as effectively as a grabby artificial surface will.

There is certainly a general impression among players that Astroturf is "faster" than grass. To my knowledge, the only published study of how turf affects speed and agility was done in 1974 by Stanitski, McMaster, and Ferguson, the same guys who measured the coefficients of restitution of various playing surfaces. Their results, for both a 40-yard straight sprint and a slalom course, are shown in **Figure 8-1** on page 238. In the latter case,

Time Advantage

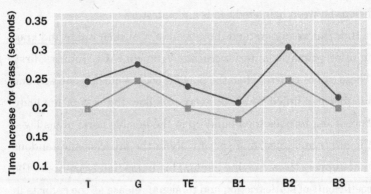

Figure 8-1. Increase in the time required to run a 40-yard straight-ahead sprint (circles) and a 40-yard slalom sprint (squares) on natural grass as opposed to Astroturf. Data are shown for a college-level tackle (T), guard (G), tight end (TE), and three backs (B). The average Astroturf time for the sprint and the slalom for all six players was 4.99 seconds and 5.17 seconds, respectively.

the players ran from the goal line to the 40 around pylons placed at the 10-, 15-, 25-, and 30-yard lines in a straight line. The conclusion we can draw from these data is that Astroturf is, in fact, faster than grass, at least the type used in these studies.

Recent data from the University of Nebraska taken by two of our team trainers, Boyd Epley and Mike Arthur, are reasonably consistent with these numbers. In comparing Astroturf with the grass on our practice fields for the 40-yard dash, they measured about a 0.2-second advantage on Astroturf for heavier players such as linemen and a few of the big backs. This advantage essentially disappears for the lighter wide receivers and defensive backs. Since we usually think about lightning speed as making the difference when it comes to the performance of running backs or the lighter pass receivers and defenders, this result tends to negate the idea that "fast" turf helps a faster team. In comparing grass and

FieldTurf, Epley and Arthur found FieldTurf to be marginally faster, perhaps by as much as 0.1 second for the heavier players.

These differences are not huge, but they can make the difference in a timing play. Using the running model we developed in chapter 1, taking a difference in 40-time to be 0.2 seconds and assuming that the acceleration boost phase on all turf surfaces is precisely 2.0 seconds, we can postulate top-end speed differences between grass and Astroturf to be 1.7 feet per second for a grass time of 4.8 seconds, or a difference of more than 5 percent in a top-end speed of 31 feet per second.

Once again, we can understand these results qualitatively by considering both the resilience and friction numbers. Astroturf and FieldTurf, with their higher coefficients of static friction, give better traction and hence the ability to accelerate faster. You might be tempted to conclude that this would allow the heavier players to accelerate faster because their weight (N) is bigger, so that the frictional traction force the ground exerts on them will be bigger as well. But remember, while the normal force is basically proportional to mass, a player's acceleration due to any given frictional force is inversely proportional to mass according to Newton's Second Law. Thus, these two effects simply cancel each other, giving no fundamental advantage to the bigger players. What we can say is that on high-friction surfaces, *all* players accelerate better. This greater traction also improves agility; one can plant a foot and change direction or speed with less danger of slipping.

Grass kills speed in two ways. Its relatively uneven surface means that a runner is constantly having to instinctively adjust his weight and muscle force to compensate for small, unexpected forces that alter his natural running motion. Its softness and tendency to absorb energy on impact mean that the runner keeps sinking in and having to pull out of the turf, losing energy with each stride. This can also be a problem with the less resilient infilled turfs. The latter effect should be more important for heavier players, who sink into the surface more easily and deeply than lighter players.

A very interesting study done at Harvard back in the mid-1970s by McMahon and Greene points to another possible advantage of Astroturf for enhancing top-end running speed. They studied speed as a function of surface resilience and had their experimental test subjects run on everything from concrete to thick, soft foam rubber. A runner's speed can be thought of as his stride length, i.e., the distance between his footfalls, divided by the time his foot is on the ground. Thus, a high running speed would correspond to a long stride accompanied by very short contact times between the runner's shoe and the running surface. What McMahon and Greene were able to show was that while stride length always increased as the track got softer, there was a range of resilience between the very soft and the very hard in which the foot-surface contact time was minimized.

This range of minimized contact time corresponds to an ideal resilience for maximizing running speed. A track designed to have a resilience in this range was constructed at Harvard. It was made of polyurethane supported on a springy wood base. Members of the Crimson track team found that their running speed was enhanced by 2 percent to 3 percent over their performance on very hard running tracks. Astroturf may offer runners the same kind of resilience advantage, while the softness of grass reduces the ratio of stride length to contact time and slows the runners. Having said all this, the several (unscientific) comparative studies of pro running back performance I've seen indicate no significant difference in either yards per carry or touchdowns per carry when grass is compared with artificial (Astroturf-style) turf. This nega-tive result could be explained by equivalent speed enhancement on the defensive side of the ball.

MISCELLANY

What other effects could a change of playing surface have on the game? To the extent that Astroturf enhances acceleration and speed, the kinetic energy and power expended in the game will in-crease. If we assume the top-end speed enhancements mentioned above, this would add another 10 percent to the nearly 100 per-cent enhancement over the past 80 years of kinetic energy gener-ated during one game (see **Figure 2-1** on page 62).

Grass and artificial turf will have different effects on punts and kickoffs that hit the ground before being touched. Wet or soft grass will tend to grab a ball and reduce the distance it will roll after coming to earth. This is primarily a result of the grass's softness; the coeffi-cient of friction is less relevant in impulsive collisions of this type.

Finally, it should be mentioned that other factors must be taken into account to completely characterize a turf's total performance profile. Water absorbency and the effect of any retained water on the various coefficients must be evaluated. In a hard freeze, these parameters can vary dramatically. (Remember the Ice Bowl!) Fields with built-in heaters can minimize the effects of ice formation, but a driving rain-storm will be hard to deal with no matter how well the field has been prepared. Ultimately, the best way to deal with such unforeseen cir-cumstances is to carry a variety of cleats in the team's equipment bag.

"MY DEFINITION OF A FAN IS THE KIND OF GUY WHO WILL SCREAM AT YOU FROM THE 60TH ROW OF THE BLEACHERS BECAUSE HE THINKS YOU MISSED A MARGINAL HOLDING CALL IN THE CENTER OF THE INTERIOR LINE, AND THEN AFTER THE GAME WON'T BE ABLE TO FIND HIS CAR IN THE PARKING LOT."

—NFL REFEREE JIM TUNNEY

CHAPTER 9

WAVES IN THE STADIUM

Now that we've fully considered the natural laws that govern how football is played and seen in detail how gear protects the players, let's examine the very real impact of one of the more powerful intangibles in determining the outcome of specific games: the fans. In this chapter we're going to talk about the physics of sound in the stadium and its role in establishing home-field advantage.

For a visiting team's offensive unit, the noise from the fans in the stands can be psychologically intimidating as well as physically challenging. Yes, these are big, strong, fearless guys, but imagine you're a quarterback and you're leading your team down the field in a last-ditch drive during a must-win game. You're down 9–14 in the fourth quarter, the 2-minute warning has come and gone, and you have no time-outs left. Worse yet, you're playing on the road in Oakland, or Kansas City, or Minnesota, which are notorious for their extra-loud fans. Now the sound level down on the field is approaching 120 decibels, the equivalent of what you'd experience in the front row of a Kiss concert—except Paul Stanley is nowhere in sight. What you can see—and, more than anything, *hear*—are 30,000 hostile fans facing you in the end zone, some of whom you know would like nothing better than to slit your throat in a dark alley. (You gotta love those Oakland fans.)

But wait, it gets worse. The guys on the other side of the line have been shifting their defensive set all afternoon, and now, with one, maybe two plays left, you can see in their eyes that they've read the play and are moving to cut off your go-to receiver's route. This means that you have to call an audible at the last second. You look down the line at your receiver and open your mouth, but you might as well be yelling into a jet engine. Now what?

THE SOUND AND THE FURY

What is sound, and how is it produced? We'll start at the end and work our way back toward the beginning. Ultimately, sound is the thing that your brain experiences when little hairs in the inner part of your ears wiggle back and forth. These hairs are attached to nerves that lead to the parts of your brain that ultimately experience the mental sensation of hearing. Why do these hairs wiggle? They wiggle in response to the motion of the fluid in your inner ear. This motion is caused in turn by your eardrum, which is basically like a piston. When the air outside your ear changes pressure, the eardrum moves back and forth, which in turn causes motion in the fluid. Why is the air changing pressure? That Oakland fan is yelling at you! Deep in his throat is his voice box, made of muscle tissue that vibrates when he's screaming. These vibrations alternately compress or rarefy the air in the voice box, creating sound waves that move up and out of his mouth. They radiate out in a roughly spherical pattern, in the same way that circular ripples are produced when a stone is dropped in water. At some point—down on the field, before each play—these waves reach your ears and you hear the abuse being heaped on you.

Let's briefly explore the analogy between water waves and sound waves. Experimenting with stones and paddles at the local pond, we notice that the crests of water waves travel at a constant speed. Sound waves have a speed, too—it's the well-known "speed of sound." When jets break the sound barrier, they are exceeding this speed. The speed of sound is determined by the speed of the individual air molecules that carry the sound waves. A local enhancement of pressure, such as that produced in the voice box, can't spread out any faster than the speed of the individual molecules that make it up.

Air molecules, in turn, have speeds determined by their kinetic energy: $(\frac{1}{64})mv^2$. This kinetic energy is proportional to their absolute temperature on the Kelvin scale (remember chapter 5). Indeed, kinetic energy is the way temperature is defined: the more kinetic energy a system of particles has, the higher its temperature. This means the speed of sound is faster in warm air than it is in cold air. In air at room temperature, the individual molecules, mostly nitrogen, move at about 1,130 feet per second, or 770 miles per hour. During the Ice Bowl, the sound speed was closer to 700 miles per hour.

A speed of 770 miles per hour may sound fast, but it's a crawl compared to the speed of light. Light travels at a speed of one billion feet per second. Light is also different from sound in another crucial way: it doesn't need a medium through which to travel. Without air, sound waves don't exist because there is nothing to compress. The same is true for water waves. But light waves, consisting of wiggling electric fields and magnetic fields, expand through space on their own.

If sound is so slow, comparatively, should a wide receiver watch the ball instead of listening to the count to determine when to start moving? After all, the center is closer to the quarterback, so he'll hear the sound coming from the quarterback's mouth first. He will then see the ball move with a time delay due only to the time required for light to travel from the (moving) ball to him. The answer is: No, it is still better for the receiver to react to the audible snap signal than to watch the ball.

Let's see if we can understand why. We'll assume that both the wide receiver and the center have the same reaction time (0.2 seconds) to an audible signal. We will also assume that the snap signal is part of a random cadence, so that the center can only react to what he hears; he can't anticipate the time he will hear the signal and react to the cadence timing. Since sound travels at about 1,100 feet per second, it will take the snap signal about 0.003 seconds, or 3 milliseconds, to reach the center. The wideout may be as far as 30 feet from the quarterback, so it could take sound up to 30 milliseconds to reach him. The reaction time of either the receiver or the center to the snap signal won't cause a reaction for another 200 milliseconds. Now light travels a foot in a billionth of a second, so it will take only a small fraction of a millisecond for the wide receiver to see that the ball has started to move after the center has reacted to the snap count. But once he sees the ball move, it will still take him 200 milliseconds to react to that visual cue.

Adding all these different times, we see that the wide receiver can start to move on the quarterback's audible signal after 230 milliseconds, whereas by watching the ball he won't begin to move until 403 milliseconds have elapsed. Thus he is significantly better off listening instead of looking. One must always keep in mind that the defensive players key on the ball anyway. The one situation when the advantage of the offense keying on the ball is significant is when it is difficult to hear the snap count because of the din of crowd noise.

MEGADETH IN THE RED ZONE

Earlier we mentioned that extreme crowd noise can reach a sound level of about 120 decibels (dB). But just what is a decibel, and how do we define sound levels in exact terms? The decibel scale was named in honor of Alexander Graham Bell, who invented the telephone. Bell also did a lot of work on the physics of sound and the human ear. Using this scale, we define the lowest sound level that the average human ear can hear to be zero dB. The term *sound level* refers to the amount of power per unit area that a sound wave delivers. Sometimes this unit is referred to as *energy flux* because it corresponds to the amount of energy that flows through a given area in a given amount of time.

What about sounds louder than 0 dB? The decibel scale is set up to be logarithmic, which means that for every increase of a given sound level by a given multiplicative factor, you add a corresponding amount to the sound level in decibels. Sounds confusing, I know, so let's see how it works using a numerical example.

Decibel Levels of Various Sounds

dB	Sound
0	threshold of hearing
10	normal breathing
20	rustling leaves
30	soft whisper
40	buzzing mosquito
50	quiet office
60	normal conversation
70	vacuum cleaner
80	busy traffic
90	Niagara Falls
100	power mower
110	heavy construction noise
120	rock concert (threshold of pain)
130	machine gun
150	jet taking off nearby
180	*Saturn* V rocket engine close by

Table 9-1. The decibel levels of various sounds and sonic environments. Notice that psychological perceptions of the level of a given sound can be deceptive, based on this scale. The sound of a normal conversation is 60 dB; sitting in front of a guitar amplifier at a Megadeth concert gets us up to 120 dB. One would be tempted to say, based on the decibel scale alone, that the concert sound level is twice that of the conversation. But clearly we think of the rock concert as being a lot louder than a conversation—certainly more than twice as loud. And it is. Using the logarithmic scale, we see that on an energy flux basis, the concert environment is actually a million times greater.

You might think that if you multiply the sound level by 10, you should multiply the decibel level by 10 as well. But if the threshold of hearing is defined to be 0 dB (see **Table 9-1**), multiplying by 10 to get the sound level would still give you 0 dB! Instead, every time you increase the sound level by a factor of 10, add 10 dB to the sound level. Thus, a sound that has 10 times the energy flux of the minimum discernible sound level has a sound level of 10 dB. Increase this by another factor of 10 and you get 20 dB. A sound level with 10,000 times greater energy flux than that at the threshold of hearing has 40 dB, and so on. Conversely, a sound level with one-tenth the energy flux of that corresponding to 0 dB has a level of −10 dB.

MAKING WAVES

The energy flux in a sound wave falls off as the square of the distance from the sound's source. This is a general property of waves, and it is easy to see why this happens. Imagine that we set up several spheres of different sizes, all with their centers at a source of sound, such as a firecracker exploding at midfield. The firecracker's explosion produces enough sound energy that the sound level at the innermost sphere with, say, a radius of 1 foot, is 100 dB.

Now, the surface area of a sphere of radius R is $4\pi R^2$. The total energy per second—or *power*—passing through this sphere is its area times the energy flux corresponding to a sound level of 100 dB. As we move to spheres of increasing radius, their surface areas increase like the square of their radii, but the total power contained in the sound wave produced by the firecracker will remain the same at any radius. In other words, as the sound wave expands outward, the amount of energy flowing through all of the spheres is the same. Since the surface area of these spheres is increasing as the square of their radii, the energy flow per second per square foot must decrease in order to keep the total power constant. Thus, the sound level drops off as $1/R^2$.

All of this means that we can now calculate how far sound carries from, say, the Patriots' Gillette Stadium during a home game. To make the calculation simple (spherical chicken mode!), think about an imaginary sphere surrounding the stadium in Foxboro with a radius of 100 yards. Now, down on the field during a defensive stand in the red zone, the sound level is, say, 120 dB. The sound level on the field is always louder than in the stands because people direct their voices in that direction. At the surface of our 100-yard-radius sphere, which is roughly at the position of one of the nosebleed seats at Gillette, the sound level will be about 100 dB.

If the sound level is 100 dB on a sphere with a radius of 100 yards centered at midfield, our rule for the rate at which energy flow drops off says that the intensity will be down by a factor of 100 at the surface of a concentric sphere with a radius of 1,000 yards—that is, with a diameter of 1.2 miles. Because a factor of 100 drop-off in sound level corresponds to a −20 dB change, when we're 0.6 miles from the stadium, the crowd noise will be 80 dB—the equivalent of the sound level of traffic on a busy street. Increasing the radius of our next sphere by a factor of 35, we're 21 miles from the stadium and in the Back Bay of Boston.

That factor of 35 drops us down roughly by 1,000 in the sound level, to 50 dB. This is a level of sound associated with normal conversation. Thus, if you're standing on Commonwealth Avenue, you theoretically should be able to hear the sound of the game. More often, though, various sorts of interference prevent this from happening. Commonwealth Avenue has an ambient noise level of at least 80 dB most of the time, which will effectively mask the noise from Foxboro. Obstacles on the ground can absorb or reflect noise, and wind can carry the sound away.

How far away from the stadium would we have to get, barring obstacles and wind, before the sound level would drop below the threshold of hearing? We'd need to drop the level in Boston, 21 miles away, by another 50 dB, or a factor of 100,000. This corresponds to another radius factor of the square root of 100,000, or about 316. According to theory, then, we could hear the noise from the crowd at a distance of about (21 × 316 miles), or more than 6,600 miles! Of course, figuring out how to get a completely still sphere of air at constant pressure 13,000 miles in diameter would be quite a trick.

TEMPEST IN A TEACUP

We know that a sound level of 120 dB is *loud*, but is there any chance that it could affect the game of football directly, such as possibly altering the trajectory of a pass?

Let's first figure out how much energy the crowd noise contains.

Consider an extreme example. The loudest noise ever recorded at a football game, according to the Royal Association for Deaf People (!), occurred in 2000 in Denver—in an undomed stadium! It had a sound level of 128.7 dB, which exceeds the threshold of pain for most people. Let's assume that Broncos fans kept the noise up at this level for a full four quarters, including halftime and quarter breaks, as well as time-outs and all other pauses in the action. This would mean nonstop yelling for almost 4 hours. Now 128.7 dB corresponds to an energy flux of 0.5 pound-force feet per second per square foot, almost a trillion times more intense than the sound of normal breathing. If we were to put a teacup full of water at midfield, how hot would the water get as a result of all this sonic energy? To calculate this, we take the energy flux and multiply it by 4 hours (14,400 seconds) and the total area surface area of the water. To make life simple, we'll just assume the water is in a cubic teacup, 3 inches on a side (0.38 square feet). (We'll make the assumption that the energy flux is the same from all directions.) This gives us a total energy of about 2,700 pounds-force feet.

This amount of energy, even expended over a period of 4 hours, will barely lift a 300-pound lineman 8 feet in the air. We can thus reasonably eliminate the possibility of sound having any major kinematic effect on the conduct of the game. As for the water in our teacup, the sound energy delivered to the cup will raise the water temperature by about 4 degrees Fahrenheit. Bottom line: sound doesn't contain much energy. This example does show, though, how incredibly sensitive our hearing is.

A COLLECTIVE WAVE

There are other waves in the football stadium. In addition to sound and light waves, there is, simply, the Wave. Yes, I'm talking about the semiorganized cheer in which fans get to their feet when they perceive the people next to them—either to the right or to the left, depending on which way the Wave is traveling—doing the same thing, and then after a moment sitting down again. Typically, this collective motion, or excitation, travels around the stadium in a circular motion, as opposed to moving in and out perpendicular to the sidelines. The group of people standing up at any one time—the crest of the Wave—is typically 10 to 20 people wide,

Fans do the Wave.

side by side, along the direction of the Wave's forward motion. Interestingly, the Wave almost always travels in a clockwise direction.

The Wave phenomenon was first noticed internationally at the 1986 World Cup soccer competition in Mexico. For this reason it is often referred to as *La Ola*, or the Mexican Wave. While American football fans sullenly argue that it was invented by cheerleaders at the University of Washington in the early 1980s, it did, in all likelihood, truly originate in Mexico, possibly as early as 1968 at the Olympic Games in Mexico City.

As good physicists interested in natural phenomena, there are several questions we should ask about the Wave. First, how does it get started? Second, what determines the speed at which it travels? Finally, why does it usually rotate in a clockwise direction about the field?

The first question appears to have been answered by three Hungarian researchers who analyzed video footage of 14 Waves in soccer stadiums containing at least 50,000 people. They were able to successfully describe the Wave using a fairly simple stimulus-reaction model of human behavior. What they discovered is that in order to initiate a Wave, at least 25 to 30 people in a cluster—a critical mass, if you will—must stand up at the same time. In their model, they assumed an "activation threshold"—basically, a probability that another person would stand to help the Wave if he or she saw people in their vicinity doing the same thing. The researchers also weighted this probability by the position of the people already standing relative to the person about to stand. The greatest impetus for standing came from people immediately to the right of the person in question at their "3 o'clock" position. This probability falls off as the angle from the right-hand direction increases (either to the front or to the back) toward the left; a person at the 9 o'clock position is least likely to cause our subject to stand. In addition, there is a weighting for distance: the farther away a person is, the less stimulus he provides to our potential stander.

This preference for right over left is inserted into the model in an ad hoc fashion to guarantee that the Wave will travel in a clockwise direction. (We are assuming that our fans are facing the field.) Why, though, does this actually happen? The researchers speculate that the visual acuity and perception of most people is weighted toward the right because most people are right-handed. This means that they will be more attuned to a person starting to stand on their right-hand side, and so they are more likely to stand in response to a person in this part of their field of view. Another possibility is that people are used to seeing the Wave go in this direction and thus are more likely to respond in the same way.

It is interesting to note that this model of Wave propagation in football fans is similar in many aspects to the propagation of sound waves. Sound is carried through the air when air molecules "communicate" with each other through collisions. This is essentially what happens with fans in the stands, except they communicate not by bumping into one another but by rising up as the Wave comes at them. They react to this visual stimulus, and their brains and leg muscles do the rest of the work.

If this similarity between the two kinds of waves is fundamental to their nature, then the Wave speed ought to depend in a similar fashion on corresponding external variables. The speed of sound, as we have discussed, depends on the air's temperature in degrees

Kelvin. As the temperature approaches absolute zero ($-491°F$), the sound speed slows to zero as well because the molecules aren't moving very fast. In other words, their communication is slowed down because it takes them longer to move over and bump into the next molecule.

In fact, the same thing does indeed appear to happen with football fans. Over the course of several years of watching the Wave at Nebraska home football games, I have noticed that there seems to be a significant dependence of Wave speed on the temperature. Plotted in **Figure 9-1** is the speed of sound in air as a function of absolute Kelvin temperature, and the speed of the Wave on the same scale. The shape of the sound-speed curve is associated with a square-root dependence on temperature. Although the uncer-

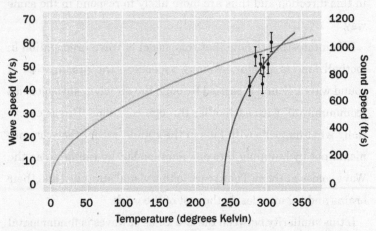

Sound Speed of Fans Doing the Wave

Figure 9-1. Data showing speed of the Wave at Nebraska's Memorial Stadium as a function of temperature. The dark gray line is a fit to the data assuming that the speed goes as the square root of the Kelvin temperature. The light gray line corresponds to the speed of sound at various temperatures. Note that absolute zero for people doing the Wave is $-40°F = -40°C = 243K$.

tainty of our individual data points is fairly large (as indicated by the vertical bars on each point), the dependence with temperature can be described qualitatively by the same kind of square-root curve. As the temperature drops, the Wave speed drops accordingly, in the same way as sound speed. This actually makes sense. If the fans are really cold and bundled up in big, heavy overcoats, they're going to find it harder to jump up when the Wave comes at them. Moreover, they'll probably be much more interested in huddling together against the cold than flailing about in the icy wind. Thus, the Wave becomes increasingly sluggish as the temperature drops.

Our graph does indicate one major difference between the two kinds of waves. Absolute zero, the temperature at which all molecular motion, and hence sound itself, stops, is −491°F. Extrapolating our data to zero Wave speed indicates an absolute zero for people at about −40°F. This makes sense too. Would you want to do the Wave at 40 below?

APPENDIX

In this appendix we will take a more in-depth look at some important physics concepts and tools, including the Cartesian grid, the Pythagorean theorem, and vectors.

To supplement our discussion of receivers' routes in chapter 3, we will specify a player's position on the football field using Cartesian coordinates. Each point on the field is assigned an x-coordinate and a y-coordinate. We'll thus designate each point like this: (x,y). In chapters 1 and 2 we considered only one kind of motion along a straight line. We thus needed only one position variable: x. The extra dimension that open-field running and passing add to the game requires an extra geometric dimension to account for the fact that the football field has both length and breadth.

In **Figure A-1** on page 260 we superimpose a Cartesian grid on the football field: yardage along the long dimension of the field is designated with the coordinate x; the yardage position across the breadth of the field is indicated by the coordinate y. Note that "x" is simply the normal yardage position that is routinely read off— "He's at the 25, 20, 10 . . . !"—except that we set its zero value to correspond to the 50-yard line. This means that negative yardage indicates 49ers' territory, and positive yardage corresponds to the Bengals' end of the field. The y coordinate will be set to zero at the center of the field as well, with y being greater than zero and increasing as we move vertically upward. As we drop below the midpoint of the field, y will have negative values.

The center of this coordinate system is called the *origin*. The origin, by definition, has the space coordinates (0,0). We can now associate with any event in a given play, such as a catch or a tackle, a unique (x,y) Cartesian coordinate pair. We will also assign a time to that event. We will usually want to reset our time clock to zero for each play, and will typically start the clock when the ball is snapped. The space coordinates will always be given in terms of yards, and the time will be given in seconds.

Let's now describe our Montana pass to Rice using this system (**Figure A-1**). Montana takes the snap at t = 0 seconds. He is at San Francisco's 18-yard line, or at −32 yards in our new coordinate system. His y position at the right hash-mark line: y = +3.1 yards. We can thus designate the snap as taking place at (−32 yards, +3.1 yards, 0 seconds) in space-time, or (−32,+3.1,0) for short. Montana drops back to the 11, rolling right 13.5 yards, sets up, and passes. This procedure takes a total of 4 seconds. Thus, the pass is launched at (−39,+16.6,+4).

In Figure A-1 our pass will be superimposed on a diagram of the football field, yard lines, the four through ten on the field is calculated with the corresponding field of position.

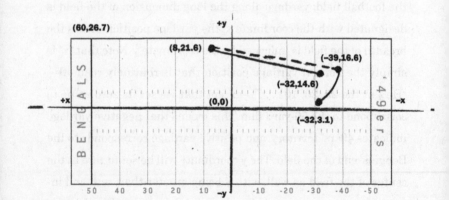

Figure A-1. Cartesian coordinate diagram of Montana-to-Rice pass good for 44 yards in Super Bowl XXIII (see page 77).

What has Rice been doing as Montana gets ready to pass to him? He has started at t = 0, spends a couple of seconds discussing the meaning of life with the Cincinnati defense, then accelerates smoothly for 2 seconds to reach his top speed of 30 feet per second. The reception point is 40 yards down field at the Bengal's 42. His lateral (y) position increases smoothly by about 7 yards over the course of his run, which puts him 5 yards from the sideline when the ball reaches him. Rice's (x,y) beginning and ending coordinates (the *space* part of space-time) are thus (−32, +14.6) and (+8,+21.6), respectively.

What are the time coordinates for the end points of Rice's run? The first one is easy: he starts to move at the snap, so it's zero. The second one is tougher. How long does it take him to reach the point of reception? We know that he accelerates from what is essentially a standstill up to 30 feet per second (10 yards per second) in 2 seconds. Thus, his acceleration over this time interval is 15 feet per second per second. We can calculate, using the techniques of chapter 1, how far he has moved after a given time interval, assuming he runs in a straight line.

Now, how far is the point of reception from Rice's starting point? Isn't it just the distance from San Francisco's 18 to the Bengals' 42 = 40 yards? Well, that *is* the distance along x that he travels, but in moving over to the flag he's traveled an additional 7 yards along y. So is his total distance 47 yards? The answer is no. In order to calculate the distance Rice actually runs, we have to make a short detour to ancient Greece and dig up an extremely useful bit of geometry: the theorem of Pythagoras.

IT'S ALL GREEK TO ME

The Pythagorean theorem can be easily understood by looking at **Figure A-2**. Consider any right triangle—one that has one of its angles equal to 90 degrees. (The sum of all three inner angles of a triangle must add up to 180 degrees.) The long side of the triangle is called the *hypotenuse*. Pythagoras tells us that the length of the hypotenuse, squared, is equal to the sum of the squares of the other two sides: $c^2 = a^2 + b^2$. For a triangle with a vertical height of 10 yards, **Table A-1** lists the distance you have to travel along the combined vertical and horizontal legs of the triangle to get from A to B, compared to the trip along the hypotenuse beginning and ending at the same points. As we can see, it is always shorter to travel along the hypotenuse. You always knew that the shortest distance between two points is a straight line, and now you know why!

More important for our West Coast offense, the Pythagorean theorem tells us how to calculate the straight-line distance between two points if we know their (x,y) coordinates. This is what we want to do with Rice's run. On a Cartesian grid, we can always view two points as being the end points of a hypotenuse. The differences in their x and y coordinates are thus simply the lengths of the two short legs of the right triangle. Given points 1 and 2, with spatial coordinates (x_1, y_1) and (x_2, y_2), respectively, the straight-line distance d_{12} between them is just the length of the hypotenuse that connects them, whose length is given by the Pythagorean theorem: $d_{12}^2 = (x_1 - x_2)^2 + (y_1 - y_2)^2$.

We're now set to figure out how long it takes Rice to get to the flag. The total distance he travels is the square root of

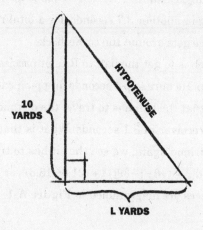

Figure A-2. Right triangle and the Pythagorean Theorem. See also Table A-1.

L	L + 10 Yards	Hypotenuse	Difference
5	15	11.2	3.8
10	20	14.1	5.9
15	25	18	7
20	30	22.4	7.6
30	40	31.6	8.4
40	50	41.2	8.8
50	60	51	9
60	70	60.8	9.2
80	90	80.6	9.4
100	110	100.5	9.5

Table A-1. Distances along the vertical plus horizontal legs of a right triangle whose height is 10 yards and whose base is the variable distance L yards. Compare with the shorter length of the hypotenuse.

$(+8-[-32])^2 + (21.6-14.6)^2$, or 40.6 yards. We can round this off to 41 yards for simplicity. This isn't much different than the x-distance that he travels: 40 yards. Rice travels 10 yards over his boost phase, so he still has 31 yards to go to the point of reception.

Now our old standby, Distance = Rate × Time is useful, and we get him to the flag in another 3.7 seconds, for a total run time of 5.1 seconds after he gets around the cornerback.

Joe Montana's job is to get the ball to Rice at precisely 7.1 seconds after the snap. He burns up 4 seconds dropping back and setting. That means that the ball has to travel the distance between quarterback and receiver in 3.1 seconds. What is that distance? Using our Greek friend again, we see that it has to travel a distance d_{12} given by $d_{12}^2 = (+8-[-39])^2 + (21.6-16.6)^2$, or 47.3 yards. All of these distances are diagrammed in **Figure A-1**.

WHAT'S YOUR VECTOR, MARVIN?

There are five vector quantities of particular relevance for the football fan: position, velocity, acceleration, momentum, and force. We have discussed all of these quantities in one dimension; their extension to vectors is easy.

Let's consider the path of a receiver running a typical circular crossing pattern (see **Figure A-3**). We'll analyze his route in terms of vectors. Our receiver—it could be standout Indianapolis Colts wide receiver Marvin Harrison—starts off from the line of scrimmage and runs straight downfield. His initial position (point 1) is indicated by a vector $\vec{r_1}$ extending from the origin (0,0) to where he stands before the ball is snapped. Position vectors give us in graphical form the same information we get by specifying a player's position in the Cartesian coordinate system as an (x,y) pair. The advantage of the vector method is that now we have a simple picture of where the player is by using an arrow, instead of having to relate an (x,y) pair to a given coordinate system.

Initially Harrison's speed, and hence the length of his velocity vector ($\vec{v}$), is small, but his acceleration ($\vec{a}$) is large. The first velocity vector off the line of scrimmage points in the direction he's running. So does his acceleration vector. The fact that the velocity and acceleration vectors are parallel means that his speed along the direction of the velocity vector is increasing; he's moving in a straight line and speeding up. The acceleration vector is just the change of the velocity vector, $\Delta\vec{v}$, over a given time interval Δt, divided by that time interval $\vec{a} = \Delta\vec{v}/\Delta t$. This equation tells us that

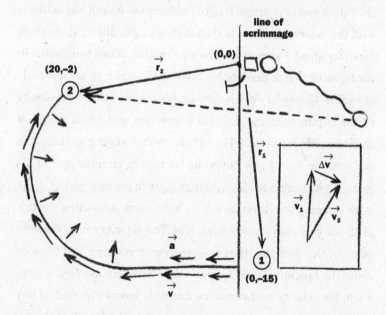

Figure A-3. A typical pass-play crossing route diagrammed in terms of vectors. The $\vec{a}$ vectors are on the side of Harrison's path toward the quarterback; the velocity vectors, $\vec{v}$, are on the other side.

the vectors $\vec{a}$ and $\vec{\Delta v}$ point in the same direction. Since both $\vec{\Delta v}$ and $\vec{v}$ are initially pointing in the same direction, our receiver continues on a straight line, with the length of his velocity vector getting continually bigger as he accelerates. Where is this acceleration coming from? We know the answer from chapter 2: from the force the ground exerts on the receiver's feet as he pushes off the line of scrimmage. If acceleration is a vector quantity, though, force has to be a vector as well: $\vec{a} = 32\vec{F}/m$. Force and acceleration always point in the same direction.

Let's return now to **Figure A-3** and our receiver. Harrison accelerates off the line of scrimmage for about 3 seconds, during which time his velocity vector gets longer and longer, while his acceleration vector remains roughly constant in length and colinear with the velocity vector. He then runs straight ahead at constant speed for about 1 second with no acceleration. When he has run 10 yards, he begins to execute his hook to the right, in a wide circle of radius 10 yards. Just before he begins his turn, his velocity vector is purely in the positive x-direction and has a length of about 8 yards per second. His velocity isn't changing, so there's no acceleration vector just before he begins the circular part of his route. As he begins his hook, things start to get interesting.

We can map out Harrison's route with a line connecting the tips of all his successive position vectors. The velocity vectors always point in the direction that the receiver is running and thus lie along the tangent line of his path. As Harrison's path curves upward, his velocity vector rotates smoothly upward as well. If he's still running at 8 yards per second, though, the *length* of the velocity vector *isn't* changing. This means, very simply, that we can change velocity without changing speed.

Now comes the weird part. According to $\vec{a} = \Delta\vec{v}/\Delta t$, if the receiver's velocity vector is changing in time, then he's accelerating. This does not necessarily mean, however, that he's speeding up or slowing down; in this case it means only that he's changing direction on his run. The acceleration vector is proportional to $\Delta\vec{v}$, with the constant of proportionality being $(1/\Delta t)$.

To understand how this works graphically, we need to learn a new trick—how to *add* vectors. Adding vectors is easy. If we have two vectors, $\vec{A}$ and $\vec{B}$, and wish to add them, we simply take the tail of $\vec{B}$ and put it at the tip of $\vec{A}$. This gives us the sum of the two vectors. Our receiver's velocity vector at some time t_2 is given by whatever his velocity vector was a bit earlier at time t_1, plus the vector associated with the change in the velocity over that time interval, $\Delta\vec{v}$: $\vec{v}_2 = \vec{v}_1 + \Delta\vec{v}$, or $\vec{v}_2 - \vec{v}_1 = \Delta\vec{v}$. We get his new velocity vector by placing the tail of $\Delta\vec{v}$ at the tip of $\vec{v}_1$. This is illustrated in the inset of **Figure A-3** on page 265. Notice, now, that $\Delta\vec{v}$ points in a direction that is roughly perpendicular to $\vec{v}_1$. Since the acceleration $\vec{a}$ is proportional to $\Delta\vec{v}$, the receiver's acceleration must be pointing toward the center of his circular path. This is in contrast with his acceleration earlier when he was moving on a straight line and speeding up. This kind of perpendicular acceleration is, in fact, typical of situations where the direction of an object's motion is changing without an attendant change in speed.

What is causing this acceleration? Newton's Second Law tells us that whenever something accelerates, there must be an unbalanced force acting on it. In this case, the unbalanced force must be acting to the right in order to cause the receiver's acceleration in that direction. This force is caused by the ground pushing on his feet in reaction to his push on the ground to his left. As long as he continues in his circular path at a constant speed, he must exert this force perpendicular to his motion. So if the ground is exerting this sideways force on the receiver, how much work is it doing on him? Answer: None! The force is not acting along his line of motion, so it can't do any work. Put another way, the receiver's kinetic energy hasn't increased because his speed hasn't changed. Therefore, no work can have been done on him.

This concept applies equally as well to the moon orbiting the earth. The moon follows an essentially circular orbital path. It is held on this path by the gravitational force exerted on it by the earth that acts along a line joining the two. This force does no work on the moon but provides just the right acceleration to keep it on its circular path.

We can also analyze Barry Sanders's juke (see page 87) using vectors and Newton's Second Law. The time required for him to make his cut is "the blink of an eye," or roughly 0.2 seconds. This means that $\vec{v_1}$ changes into $\vec{v_2}$ in this time. We can graphically calculate the change in velocity during this time, $\vec{\Delta v}$, by remembering that $\vec{\Delta v} = \vec{v_2} - \vec{v_1}$. Then we'll get $\vec{a}$ from $\vec{\Delta v}$ using $\vec{a} = \vec{\Delta v}/\Delta t$. (Remember $\vec{a}$ and $\vec{\Delta v}$ point in the same direction.) Now, to get $\vec{\Delta v}$ we need to *subtract* $\vec{v_1}$ from $\vec{v_2}$. But wait a minute—we've never subtracted vectors before, we've only added them! Don't worry. Subtracting $\vec{v_1}$ from $\vec{v_2}$ is exactly the same thing as adding $(-\vec{v_1})$ to $\vec{v_2}$. But what does $-\vec{v_1}$ look like? Simple. It has the same length as $\vec{v_1}$ but it points in the opposite direction. Now we just use our rule of vector addition to get $\vec{\Delta v}$; in the juke we looked at, $\vec{\Delta v}/\Delta t = 4.0$ g's.

NOTES

For general reference, *Total Football II: The Official Encyclopedia of the National Football League*, edited by Bob Carroll, Michael Gershman, David Neft, and John Thorn, with statistics provided by the Elias Sports Bureau (HarperCollins, New York, 1999), proved invaluable.

I also found *Joe Montana's Art and Magic of Quarterbacking*, by Joe Montana with Richard Weiner, *Sports Illustrated*'s *Football: A History of the Professional Game* by Peter King, and the series of books on football by John Madden to be particularly useful.

INTRODUCTION

"Tell them you touched it!"   details of the Immaculate Reception, P. King, *Football: A History of the Professional Game* (Bishop, New York, 1997), p. 64.

CHAPTER 2

footing unreliable at best   details of the Ice Bowl, L. Krantz, *Not Till the Fat Lady Sings* (Triumph, Chicago, 2003), p. 18.

He told Starr to call a 30 Wedge.   P. King, *Football: A History of the Professional Game* (Bishop, New York, 1997), p. 56.

assume the reaction time of most players to be about 0.2 seconds
"Evaluation of oculomotor response in relationship to sports performance," G. Harbin, L. Durst, and D. Harbin, *Med. Sci. Sports Exer.* **21**, 258 (1989).

studies of human running   See, e.g., "A theory of competitive running," J. B. Keller, *Phys. Today* **26**, 42 (1973).

how the size and speed of linemen in the NFL have changed since 1920
"Changes in body size of offensive players in the National Football League: a 76-year review of 27,744 players," F. I. Katch and K. D. Monahan, *J. Am. Coll. Sports Med.* **30**, S239, Abstract 1359 (1988).

haven't had a big effect on other types of major injuries "A multivariate risk analysis of selected playing surfaces in the National Football League, 1980 to 1989: An epidemiologic study of knee injuries," J. W. Powell and M. Schootman, *Am. J. Sports Med.* **20**, 686 (1992).

"Living with artificial turf: a knowledge update. Part 1: basic science," I. M. Levy, M. L. Skovron, and J. Agel, *Am. J. Sports Med.* **18**, 406 (1990).

"Living with artificial turf: a knowledge update. Part 2: epidemiology," M. L. Skovron, I. M. Levy, and J. Agel, *Am. J. Sports Med.* **18**, 510 (1990).

the way, say, defensive backs do "Life in the trenches," P. Barber, in *Total Football II*, B. Carroll, M. Gershman, D. Neft, and J. Thorn eds. (Harper Collins, New York, 1999).

his body temperature at a normal 98.6°F D. C. Nieman, *Fitness and Sports Medicine: An Introduction*, chapter 11 (Bull, Palo Alto, 1990).

consume more than 10,000 food calories without gaining an ounce
Ibid, chapter 9.

CHAPTER 3

the six most important yards in 49ers history *Joe Montana's Art and Magic of Quarterbacking*, J. Montana with R. Weiner (Henry Holt, New York, 1997).

past the line of scrimmage at midfield "Pursuit and evasion strategies in football," J. O'Connell, *Phys. Teach.* **33**, 516 (1995).

past the line of scrimmage at midfield "The linebacker's problem," L. B. Anderson, Institute for Defense Analysis (ret.), private communication with the author.

past the line of scrimmage at midfield "Chase problems," S. R. Dunbar, *UMAP Journal* V **15**, 351 (1994).

where L is the initial straight-line distance between the two players "A football chase on video," N. F. Derby, R. G. Fuller, and T. A. Summers, *Phys. Teach.* **35**, 359 (1997).

CHAPTER 4

weak in the velocity ranges of relevance to football See, e.g., *Elementary Mechanics of Fluids*, H. Rouse (Wiley, New York, 1946).

weak in the velocity ranges of relevance to football "The drag force on an American football," R. G. Watts and G. Moore, *Am. J. Phys.* **71**, 791 (2003).

This larger area turns out to be 0.41 feet squared "The physics of kicking a football," P. J. Brancazio, *Phys. Teach.* **23**, 403 (1985).

at least two research groups; more recent and extensively published results "Wind-tunnel measurements of the aerodynamic loads on an American football," W. J. Rae and R. J. Streit, *Sports Eng.* **5**, 165 (2002).

kickoff distances in the home stadiums of all the teams Denver played in the 2001 and 2002 seasons Data obtained from the NFL's Web site, www.nfl.com.stats.

CHAPTER 5

a leg mass (including the foot) of about 35 pounds The Biomechanics of *Sports* (2nd ed.), J. G. Hay (Prentice-Hall, Englewood Cliffs, NJ, 1978),p. 136.

relative to the hip joint just before contact with the ball; the foot is in contact with the ball for about 8 milliseconds "The curve kick of a football I: impact with the foot," T. Asai, M. J. Carré, T. Akatsuka, and S. J. Haake, *Sports Eng.* **5**, 183 (2002).

about 1 inch, just below the ball's center Play Football the NFL Way, T. Bass (St. Martin's Griffin, New York, 1990), p. 367.

0.050 lbm-ft² for the long axis "Rigid-body dynamics of a football," P. J. Brancazio, *Am. J. Phys.* **55**, 415 (1987).

same way a knuckleball does when thrown by a baseball pitcher The Physics of Baseball, 3rd ed., R. K. Adair (HarperCollins Perennial, New York, 2002).

several nice articles on the subject; "kicker's dilemma" "The physics of kicking a football," P. J. Brancazio, *Phys. Teach.* **23**, 403 (1985).

a punter from Auburn, Terry Daniel, to test the idea Sports Illustrated, 8 November 1993, p. 143.

increase the opening angle of the goalposts for his kicker "How to kick a field goal," Daniel C. Isaksen, *College Mathematics Journal* **27** (1996), 267.

an article on methods kickers use to doctor balls Sports Illustrated, 4 October 1999, p. 57.

CHAPTER 6

this kind of punt will do one of three things; no matter how the ball is launched; high punts that are wobbly (or tight) spirals "Rigid-body dynamics of a football," P. J. Brancazio, *Am. J. Phys.* **55**, 415 (1987).

the published reports on their work "Why does a football keep its axis pointing along its trajectory?" P. J. Brancazio, *Phys. Teach.* **23**, 571 (1985).

the published reports on their work "A geometric theory of rapidly spinning tops, tippe tops, and footballs," H. Soodak, *Am. J. Phys.* **70**, 815 (2002).

when a southpaw passes, it veers left; the vicinity of 600 revolutions per minute; vacuum value of 1.67 to about 1.9 "Flight dynamics of an American football in a forward pass," W. J. Rae, *Sports Eng.* **6**, 149 (2003).

when a southpaw passes, it veers left "Mechanics of the forward pass," W. J. Rae, in *Biomedical Engineering Principles of Sports*, ed. G. K. Hung (Kluwer/Plenum, New York, 2003).

when a southpaw passes, it veers left Quarterbacking, Bart Starr with Mark Cox (Prentice-Hall, Englewood Cliffs, New Jersey, 1967), 168–69.

curving action of baseballs The Physics of Baseball, 3rd ed., R. K. Adair (HarperCollins, New York, 2002).

curving action . . . of soccer balls "The curve kick of a football II: flight through the air," M. J. Carré, T. Asai, T. Akatsuka, and S. J. Haake, *Sports Eng.* **5**, 193 (2002).

Professor Marianne Breinig at the University of Tennessee more details of this can be found at http://footballphysics.utk.edu.

CHAPTER 7

developments of the early 1940s "Tools of the Trade," B. Riffenburgh, in *Total Football II: The Official Encyclopedia of the National Football League*, B. Carroll, M. Gershman, D. Neft, and J. Thorn, eds. (HarperCollins, New York, 1999).

CHAPTER 8

According to a report "Report on Veterans Stadium Turf Replacement," Jonathan A. Saidel, Office of the Controller, City of Philadelphia, October 15, 2001.

injuries that might befall football players; one study of this type published; restitution of various playing surfaces "Synthetic turf and grass: a comparative study," C. L. Stanitski, J. H. McMaster, and R. J. Ferguson, *Am. J. Sports Med.* **2**, 22 (1974).

more likely to cause serious injury than playing on natural grass; on either grass or infilled turfs See, e.g., "Is there a relationship between ground and climatic conditions and injuries in football?" J. Orchard, *Sports Med.* **32**, 419 (2002), and references therein.

more likely to cause serious injury than playing on natural grass; on either grass or infilled turfs "A multivariate risk analysis of selected playing surfaces in the National Football League: 1980 to 1989," J. W. Powell and M. Schootman, *Am. J. Sports Med.* **20**, 686 (1992).

more likely to cause serious injury than playing on natural grass; on either grass or infilled turfs "Living with artificial grass: a knowledge update; Part I: basic science," M. Levy, M. L. Skovron, and J. Agel, *Am. J. Sports*

Med. **18**, 406 (1990). "Living with artificial grass: a knowledge update; Part II: epidemiology," M. L. Skovron, M. Levy, and J. Agel, *Am. J. Sports Med.* **18**, 510 (1990).

from a height of 18 inches; the age and condition of the field "The validity and relevance of tests used for the assessment of sports surfaces," B. M. Nigg, *Med. Sci. Sports Exer.* **22**, 131 (1990).

from a height of 18 inches; the age and condition of the field "Turf Going," K. Newell, *Coach Ath. Director* **73**, 56 (November 2003).

from a height of 18 inches ASTM Standard Designations F 355-95; F 1702-96; F 1936-98.

a G-max approached . . . 1,300; the age and condition of the field; different types of shoes "Evaluation of playing surface characteristics of various in-filled systems," A. S. McNitt and D. Petrunak, preliminary report from the Department of Crop and Soil Sciences, Pennsylvania State University (2003); http://cropsoil.psu.edu/mcnitt.

proportional to each other; different types of shoes "Differences in friction and torsional resistance in athletic shoe–turf surface interfaces," R. S. Heidt, S. G. Dormer, P. W. Cawley, P. E. Scranton Jr., G. Losse, and M. Howard, *Am. J. Sports Med.* **24**, 834 (1996).

" . . . if conditions change."; on either grass or infilled turfs "Athletic Field Surfaces: new products, old questions," L. Schnirring, *Phys. Sports Med.*, Oct. 1, 1999.

different types of shoes "Shoe-surface traction of conventional and in-filled synthetic turf football surfaces," M. Shorten, B. Hudson, and J. Himmelsbach, Proceedings of the XIX International Congress of Biomechanics (ed P. Milburn et al.), University of Otago, Dunedin, New Zealand (2003). Available at http://biomechanica.com.

different types of shoes "Torques developed by different types of shoes on various playing surfaces," R. W. Bonstigl, C. A. Morehouse, and B. W. Niebel, *Med. Sci. Sports* **7**, 127 (1975).

different types of shoes "Cleat-surface friction on old and new Astroturf," K. D. Bowers Jr. and R. B. Martin, *Med. Sci. Sports* **7**, 132 (1975).

different types of shoes; on either grass or infilled turfs; infilled surfaces are somewhat safer than grass "Football cleat design and its effects on anterior cruciate ligament injuries," R. B. Lambson, B. S. Barnhill, and R. W. Higgins, *Am. J. Sports Med.* **24**, 155 (1996).

different types of shoes; any value of $\mu_{static} > 0.4$ *is dangerous* "The shoe-surface interface and its relationship to football knee injuries," J. S. Torg, T. C. Quedenfeld, and S. Landau, *J. Sports Med.* **2**, 261 (1974).

but this is not a very big effect "The effect of ambient temperature on the shoe-surface interface release coefficient," J. S. Torg, *Am. J. Sports Med.* **24**, 29 (1996).

on either grass or infilled turfs "National Football League 1974 Injury Study," Stanford Research Institute Report (1974).

on either grass or infilled turfs "Injury risk in men's Canada West University football," B. E. Hagel, G. H. Fick, and W. H. Meeuwisse, *Am. J. Epidem.* **157**, 825 (2003).

(wet versus dry) are varied "Report on Injury Surveillance System," National Collegiate Athletic Association (1993).

infilled surfaces are somewhat safer than grass Bill S. Barnhill, M.D., Panhandle Sports Medicine, private communication.

between helmeted heads and Astroturf "Epidemiology of concussion in collegiate and high school football players," K. M. Guskiewicz and N. L. Weaver, *Am. J. Sports Med.* **28**, 643 (2000).

enhancing top-end running speed "Fast Running tracks," T. A. McMahon and P. R. Greene, *Scientific American* **239**, 148 (1978).

CHAPTER 9

the Royal Association for Deaf People www.royaldeaf.org.uk. See also *The Guinness Book of World Records.*

the Wave almost always travels in a clockwise direction; three Hungarian researchers who analyzed video footage I. Farkas, D. Helbing, and T. Vicsek, *Nature* **419**, 131 (2002).

ACKNOWLEDGMENTS

There are five people who contributed in unique and critical ways to the writing of this book. The first is Jeff Schmahl, the former head of Huskervision at the University of Nebraska, who conceived of the idea of 1-minute physics lessons for the Husker fans at Nebraska home games. He gave me my first important lessons in communicating physics ideas to the public. These lessons were continued by Brad Minerd, who pushed the Football Physics project at NFL Films and was able to make even me look good on 16mm film. Chris Potash, my editor for the first edition of this book, has furthered my education. My original concept, "a physics text in disguise," was dropped in the early going, to be replaced by his more enlightened vision: "enriching reading for students of the game." Chris alternatively held my hand and hit me over the head with a plank when I needed it. A true student of the game with a voluminous knowledge of its history, Steve Friedman, stepped in at critical junctures to provide me with "football color" to illustrate various aspects of the narrative. Finally, Dr. Cliff Bettis, my colleague in the Department of Physics and Astronomy at the University of Nebraska, has given me countless ideas for interesting topics to discuss in relation to the general topic of football physics, and interesting ways to present these topics. The original lectures I presented to the Husker fans in Memorial Stadium were, in many ways, just an excuse to show people really cool physics demonstrations. The ideas for many of these topics and demonstrations came directly from Cliff. I owe these five guys big-time.

The physics of baseball, golf, and soccer have all been the subject of excellent books; tennis and track have received extensive discussion in various technical journals. The reason why football has been largely ignored escapes me. It is, after all, the best game there is. Having said this, there are two scientists who stand out for their groundbreaking work in this field: Bill Rae, an aeronautical engineer at the University of Buffalo, and Peter Brancazio, professor emeritus of physics at Brooklyn College. Both have generously discussed their work with me. A number of physicists, mathematicians, and engineers have given me good ideas and useful council on this topic. They are my colleagues Dan Claes, Steve Dunbar, Roch Gaussoin, Davis Marx, and Ted Jorgenson at Nebraska; Marianne Breinig of the University of Tennessee; Don Sparlin of the University of Missouri at Rolla; William Moss of the Lawrence Livermore National Laboratory; Daniel Isaksen of the University of Notre Dame; Robert Watts of Tulane University; and Robert Adair of Yale University.

People who make a living in the game of football have been invaluable in terms of providing insights and ideas. They are also more interesting to talk to than physicists. At Nebraska, I would particularly like to thank Boyd Epley, John Ingram, Bryan Bailey, Mike Arthur, Scott Downing, and Frank Solich for useful discussions. In the pro ranks, Ernie Adams of the New England Patriots and punter Louie Aguiar have been especially helpful.

I have benefited greatly from my association with a number of media producers and "creative people," including Kay Dowd, Tonya McMillion, Leslie Waechter, all student broadcasting majors and interns at Huskervision; Kirk Hartman, the current creative director of Huskervision; and Matt Miller and Eternal Polk at NFL Films. On the other side of the coin, Kelly Bartling and Randy Atkins taught me a great deal about dealing with the media.

What use is a professor without students? A large number of them at Nebraska have made an invaluable contribution to this work. Student-athletes Josh Brown, Kyle Larson, and David Dyches have given me much insight into the science and psychology of kicking and punting. The others are Tony Caruso, Maya Fabrikant, Glenn Gronniger, Brandon Jordan-Thaden, Nick Loomis, Gary Pike, Nick Reding, and Justin Zohner. Luke Reinbolt deserves special mention in this regard, for both his athletic and his statistical prowess.

Finally, I would like to thank Charles Coren, Mike Lubell, Mary Hynes-Berry, Gordon Berry, Gerald Wasserburg, Ed Hammond, Jimmy Vines, Todd Chromzak, Dan Lasman, and Bert, Fred, Chris, Annabeth, and William Gay, for various contributions too numerous to mention.

Timothy Gay
Lincoln, Nebraska
May 2004

INDEX

Boldface page references indicate illustrations and photographs.

A

Acceleration
  boost phase, 49, 50, 64–65, 74, 239
  change in collisions, 27–30
  cruise phase, 49, 50
  due to gravity, 30–31
  effect of friction on, 239
  force needed for, 50
  G-max value, 231
  of linemen off the line, 48–52
  perpendicular, 267–68
  quickness, 74
  vectors, 86, **265**, 265–67
Accuracy
  kick, 156–62
  pass, 79–81
Adams, Ernie, 197–98
Aguiar, Louie, 120
Aikman, Troy, 171
Air density, 110–15
Air drag
  cause of, 109
  effect of, on
    air density, 111
    altitude, 115
    area of ball, 111
    drag coefficient, 110, 111–12
    humidity, 126–27
    temperature, 127
    wind, 123–25
  effect on
    hang time, 116–17, 118–19
    helium-filled football, 154–56

kinetic energy, 117
nonrotating ball, 147
pass, 171, 178, 181, **182**,
  183–84, 189–92, **192**
punts, 197
range, **104**, 118–19
speed, 102
torque, 178, 181, **182**, 183–84
Magnus force, 193–95
relationship to velocity, 109
skydiving example, 109–10
yaw angle/force, 191, **192**,
  193–94
Allen, Bryan, 63
Altitude
effect on
  air density, 113–15, 116
  hang time, 116–17
  range, 118, 119–23, **121–22**
  timing, 118
of NFL team stadiums, 123
Anderson, Ken, 70
Angle-of-attack drag force, 189–90
Angular accuracy, 158–59
Angular momentum, 132–33
  of bicycle, 178–79
  conservation of, 132–33, 146,
    178, 188
  effect of frictional forces on, 185,
    186, 187
  gyroscope and, 179
  kicking and, 132–33, 135–36,
    142, 144–47
  of merry-go-round, 130

279